75 Easy Physics Demonstrations

Thomas Kardos
illustrated by Nicholas Soloway

Dedication

This book is dedicated to my darling wife, Pearl, who throughout this project assisted me with great patience. As a nonscience educator, she helped me develop this book into an easy-to-use and comprehensible resource.

1 2 3 4 5 6 7 8 9 10

ISBN 0-8251-2806-4

J. Weston Walch, Publisher
P. O. Box 658 • Portland, Maine 04104-0658

Printed in the United States of America

Contents

Preface ... *vii*

My Philosophy of Education ... *viii*

Suggestions for Teachers ... *ix*

Equipment ... *x*

Safety Procedure ... *xi*

1. Physical Change and Properties of Matter ... 1
2. Energy Waves and Energy Forms ... 3
3. Energy Is Not Matter ... 5
4. Energy (Heat) Expands Matter ... 6
5. Absorption of Heat ... 8
6. Radiant Energy ... 9
7. Vacuum Bottles ... 9
8. Kinetic Molecular Theory: States of Matter ... 10
9. Pressure of Air ... 12
10. Air Pressure and Soda Can ... 13
11. Work from Air Pressure ... 13
12. Light Travels in a Straight Line ... 14
13. Pinhole Camera ... 15
14. Angle of Incidence Equals Angle of Reflection ... 16
15. Mirror Images: Depth ... 17
16. Bending Light: Laws of Refraction ... 18
17. Motion Picture Effect ... 19
18. Electromagnetic Spectrum ... 20
19. Frequency, Wavelength, and Amplitude ... 21
20. The Light Spectrum: Color of Objects ... 22
21. Laser Light ... 23
22. Lenses ... 24
23. Refraction of Light #1 ... 24
24. Refraction of Light #2 ... 25
25. Breaking Light Apart ... 25
26. Potential Energy #1 ... 26
27. Potential Energy #2 ... 27
28. Speed, Velocity, and Friction ... 27

29. Acceleration 28
30. Bernoulli's Principle 29
31. Newton's Third Law of Motion 31
32. Magnets and Poles 32
33. Making a Magnet 33
34. Inducing Magnetism 33
35. Magnetic Fields 34
36. Magnetic Compass 35
37. Destroying a Magnet 36
38. Static Electricity 37
39. Static Electricity: Van de Graaff Generator 38
40. Electricity Makes a Magnet 39
41. Electromagnets 40
42. Making Electromagnets Stronger 41
43. Inducing Electricity by Magnetism 42
44. Inducing Electricity: Electromagnets 43
45. Series Circuits 44
46. Parallel Circuits 45
47. Short Circuits 46
48. Measuring Voltage 47
49. Measuring Current 47
50. Measuring Resistance 48
51. Ohm's Law 49
52. Thermometers and Temperature 50
53. Calories/Calorimeter 51
54. Force: Measuring Force 52
55. Work: Measuring Work 53
56. Measuring Friction 54
57. Center of Gravity 55
58. Lever: Mechanical Advantage 56
59. Wheel and Axle 57
60. Teeter-Totter: Moment 58
61. The Inclined Plane 59
62. Pulleys 60
63. Friction and Machines 61
64. Gravity 62
65. Acceleration by Gravity 63
66. Weight 64
67. Pressure 65
68. Measuring Pressure 66
69. Density and Specific Gravity 67

70. Archimedes' Principle 69
71. Making Sounds 70
72. Loudness and Sound (Decibels and Amplitude) 71
73. Pitch (Frequency) and Sound 73
74. Transmission of Sound Through Materials 74
75. Resonance 75

Appendix *77*

1. Temperature Conversion (Celsius to Fahrenheit) 78
2. Temperature Conversion (Fahrenheit to Celsius) 79
3. Melting and Boiling Points of Elements 80
4. Range of Resistances 81
5. Coefficients of Volume Expansion 81
6. Electrochemical Equivalents 81
7. Wavelengths of Various Radiations 81
8. Specific Heat of Materials 82
9. Coefficient of Linear Expansion 82
10. Density of Liquids 83
11. Altitude, Barometer, and Boiling Point 83
12. Specific Gravity 83
13. Units: Conversions and Constants 84–99

Index *101*

Preface

As a middle school teacher, many times I found myself wishing for a quick and easy demonstration to illustrate a word, a concept, or a principle in science. Also I often wanted a brief explanation to conveniently review basics and additional information without going to many texts.

This book is a collection of many classroom demonstrations. Explanation is provided so that you can quickly review key concepts. Basic scientific ideas are hard to present on a concrete level; this book fills that specific need. Some more challenging demonstrations have been included to assist you with more advanced students. Some of the demonstrations can be repeated at home and also can be adapted by you as student laboratory activities.

The actual teacher demonstration is something full of joy and expectation, like a thriller with an unexpected twist ending. Keep it that way and enjoy it! Try everything beforehand.

We need to support each other and leave footprints in the sands of time. Teaching is a living art. Happy journey! Happy sciencing!

—Thomas Kardos

My Philosophy of Education

My philosophy of education involves many themes:

1. Students must feel like participants in the joy of sciencing.
2. Students need many hands-on experiences. Invite them to experiment at home. This invitation must be limited by the availability of equipment and the relative safety of the activity. Experiments can be done with much informal equipment such as recycled jars, soda cans, and bottles. Inexpensive plastic measuring cups can replace graduated cylinders.
3. Students need to form in their own minds a concept of what science is. Do not encourage rote memorization. Science is a series of stories that need the participant's intervention. Let your students jump in and get involved in these stories.
4. Teachers do not have to answer all student questions. It is wonderful to let your students know that you are a limited resource. Let students go out and find some difficult answers. Maybe there are none. Nobody on this planet has all the answers. It is important that you teach your students the concept that humans have limits, but these can change. Let students know that through networking (cooperative effort) they, too, can find some of the harder answers.
5. People are concrete operators. Their learning starts with real objects and lots of manipulations and eventually ends in abstract reasoning and concept formation. This is why people draw sketches for you, to explain their ideas. Read summaries or explanations of Piaget's learning theory; it will change your teaching style for life.
6. Be open to change! Be prepared to change as you progress in teaching. The world around us changes and so must our teaching style.
7. Finally, realize that you cannot do it all. Your many science students become your followers. You will start a science explosion! This is your real opportunity!

Suggestions for Teachers

1. A • (bullet) denotes a demonstration. Several headings have multiple demonstrations.

2. MATERIALS: Provides an accurate list of materials needed. You can make substitutions and changes, as you find appropriate.

3. Since many demonstrations are small and are not clearly visible from the back of the room, you will need to take this into account as part of your classroom management technique. Students need to see the entire procedure, step by step.

4. Some demonstrations require that students make observations over a short period of time. It is important that students observe the changes in progress. One choice is to videotape the event and replay it several times.

5. Some demonstrations can be enhanced by bottom illumination: Place the demonstration on the overhead projector and lower the mirror so that no image is projected overhead.

6. Encourage students to repeat certain carefully selected demonstrations in class or at home.

7. Key words are included in the Index for easier access to the demonstrations.

8. I use a 30-cup coffeepot to heat water for student experiments and to perform many demonstrations in lieu of an electric hot plate, pans, and more cumbersome equipment.

9. I may use temperature Fahrenheit in some places since the majority of younger students relate to it better.

10. Just a few demonstrations may appear difficult to set up, for they have many parts. Be patient, follow the listing's steps, and you will really succeed with them.

EQUIPMENT

- Sometimes, though rarely, I will call for equipment that you may not have. An increasing growth in technology tends to complicate matters. Skip these few demonstrations or borrow the equipment from your local high school teacher. Review with him or her the proper and safe use of it. These special demonstrations will add immensely to your power as an effective educator and will enhance your professionalism.
- Try all demonstrations in advance, to smooth your show. If something fails, enjoy it and teach with it. Many great science discoveries had to be done over many times before their first success. Dr. Land had to do more than 11,000 experiments to develop the instant color photograph. Most people would have quit long before that.
- One of my favorite techniques is to use a camcorder and show the demonstration on a large monitor.

SAFETY PROCEDURES

- Follow all local, state, and federal safety procedures: Protect your students and yourself from harm.
- Attend safety classes to be up-to-date on the latest in classroom safety procedures. Much new legislation has been adopted in the recent past.
- Have evacuation plans clearly posted, planned, and actually tested.
- Have an ABC-rated fire extinguisher on hand at all times.
- Learn how to use a fire extinguisher properly.
- Label all containers and use original containers.
- Wear required safety equipment when handling hazardous materials, such as laboratory acids or anything stronger than ordinary vinegar.
- Practice your demonstration if it is totally new to you. A few demonstrations do require some prior practice.
- Keep demonstration at a distance so that no one is harmed should anything go wrong.
- Have students wash their hands whenever they come in contact with anything that may be remotely harmful to them, even if years later, like lead.
- Neutralize all acids and bases prior to disposal.
- Dispose of demonstration materials in a safe way. Obtain your district's guidelines on this matter.

DISCLAIMER

The safety rules are provided only as a guide. They are neither complete nor totally inclusive. The publisher and the author do not assume any responsibility for actions or consequences in following instructions provided in this book.

1. PHYSICAL CHANGES AND PROPERTIES OF MATTER

A **physical change** occurs when a material alters in size or shape but still remains the same material.

MATERIALS: piece of paper, a rubber band, a small piece of wood (splint)

- Take a piece of paper and tear it. It is still paper. Take a rubber band and stretch it. Take a small piece of wood and break it.

Material objects have many physical **properties**. Some properties are listed here with examples of attributes that describe them. Your students may benefit by using this list when describing objects. By repeating this activity many times in a year, students become better observers. This section is provided only as a quick reference.

TABLE: PROPERTIES OF MATERIALS

PROPERTY	**EXAMPLE**
Color	Blue, red, green
State	Solid, liquid, gas, (plasma)
Measurements	Length, width, height, mass, volume
Material	Paper, wood, metal, plastic
Temperature	How hot is it?
Quantity	How many are there?
Transparency	Can you see through it?
Magnetism	Does a magnet attract it?
Electrical conductivity	Conductor, semiconductor, or insulator
Buoyancy	Does it float?
Sound	Does it make noises on standing?
Edibility	Can you eat it?
Odor	Does it smell?
Fragility	Does it break easily?
Hollowness	Is it empty inside?
Fluorescence	Does it glow under black (UV) light?
Layers	Is it made in layers?
Moisture	Is it wet, dry?
Flexibility	Can you bend it?
Shine	Does it have a luster, or is it dull?

Hardness	Firm, solid, soft, spongy
Texture	Rough, bumpy, smooth
Holes	Does it have holes? where? how many?
Solvency	Does it dissolve materials? what kinds of materials?
Solubility	Does it become dissolved when placed in liquids?
Acid material	Check with litmus paper: both red
Base material	Check with litmus paper: both blue
Neutral material	Neither litmus changes colors
Schlieren	Does it produce stringers of coloring (schlieren)? many stringers when mixed, and appears in suspension; after a while, a mixture results.

- Display an array of objects and describe them in terms of one or several of these properties.

On a higher level, some physical properties can be described as follows.

TABLE: PHYSICAL PROPERTIES OF MATTER

PHYSICAL PROPERTY	**UNIT**
Density	10^{3}g cm^{-3}
Young's modulus	10^{10} N m^{-2}
Specific heat capacity	J kg^{-1} K^{-1}
Specific latent heat of fusion	10^{4} J kg^{-1}
Thermal conductivity	W m^{-1} K^{-1}
Resistivity	$10^{-8}\rho$ m
Linear expansivity	10^{-6} K^{-1}

Density: mass per unit of volume. Without units, it is specific gravity, with water set as 1. Objects with S.G. <1 float, objects with S.G. >1 sink.

Young's modulus: deals with the elasticity of a material and how much it can be stretched before it deforms.

Specific heat: deals with the speed at which heat energy can be absorbed and given off.

Heat of fusion: deals with the amount of heat energy needed to change the state of the material (e.g., solid to liquid), different than in just normal heating. Each material has its own heat of fusion.

Thermal conductivity: the amount of heat energy that a specific substance can conduct.

Resistivity: the resistance to conducting an electric current, unique to each material and low in most metals.

Linear expansivity: the amount of linear expansion when heat is applied.

NOTE: These variables change with temperature.

2. Energy Waves and Energy Forms

Energy travels in waves. Waves are either transverse or longitudinal. **Transverse waves** vibrate (oscillate) at right angles to the direction of the wave. When two waves travel in opposite directions, they "buck" or oppose each other. When two waves travel in the same direction, they "aid" or add their energies together.

MATERIALS: about 10 to 15 feet of rope or string, or a Slinky™ (the metal ones work best)

- Have two people hold a rope about 8–12 feet apart, and ask one of them to swing the rope gently up and down on one end. Observe how the energy of small swings is transmitted to the other end in a wavelike motion.

- Have two people take a Slinky and pull it apart about 15 feet. Let them wave it up and down gently; observe the energy waves. Have them start a wave on each end simultaneously, so the class can observe what happens when two waves oppose (buck) each other. Ask one of them to send several waves away from the same end, and observe how waves help (aid) each other.

Longitudinal waves vibrate (oscillate) in the direction of the wave. An example are sound waves.

MATERIALS: Slinky (the metal ones work best)

- Stretch a Slinky about 15 to 20 feet. Have a student hold the other end securely. Wait till the spring stops moving. Carefully squeeze two or three spring loops together and watch the "squeeze" travel forth and back along Slinky. These are longitudinal waves. In the places where the springs are closer than normal, you have wave compression. In the places where the springs are fully apart, you have wave rarefaction.

- While holding the Slinky ready, as in the previous demonstration, touch the end of it to a student's ear. Generate several longitudinal waves. The student will be able to hear the vibrations of the spring. The vibrations will be similar to the sounds of a synthesizer.

Following are some forms of energy.

TABLE: ENERGY FORMS

FORM	EXPLANATION
Sound	Waves in air caused by vibrations, such as horn blowing, objects hitting, talk, bark of dog, etc.
Electrical	Natural and man-made energy form; travels in wires; convenient to use.
Chemical	Stored in plants, animals, and minerals; digestion releases food energy.
Atomic	Energy stored in the nucleus of the atom; used to make electricity; the sun is a star that is a nuclear furnace in progress.
Light	A form of energy produced by the sun and lamps, visible to the human eye.
Heat	Causes molecules to move; produced by human digestion and changes from other energy forms.
Mechanical	The energy in the motion of objects, such as moving machines, falling objects, etc.

MATERIALS: a small light with a bulb, a flashlight, baking soda, vinegar, a glass with some water, a teaspoon, keys on a ring

- Turn the small light on. Observe the **light energy**. Mention that after a few minutes the bulb will be too hot to touch. This invisible radiant energy is **heat energy** (infrared).

- Place a couple of teaspoons or a tablespoon of baking soda in a glass half-filled with water and stir it. Add some vinegar to this mixture. A fairly quick and violent **chemical reaction** will result. (Do it over a sink.)

- Turn on the flashlight. The **chemical energy** from the reaction in the battery makes electricity for the flashlight.

- Make a sound by jingling your keys or tapping your desk with another object. The vibrations you make will result in **sound energy** that can be heard.

3. Energy Is Not Matter

These demonstrations will clarify that matter and energy are not one and the same.

Materials: a flashlight, a scale, a glass full of water

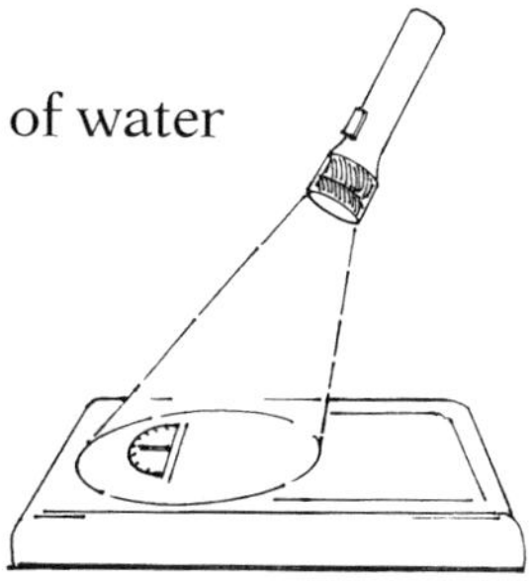

- Take a flashlight and shine it on a bathroom scale. Notice that when the beam strikes the scale, the scale shows no increase in weight. **Energy has no mass.**

- Shine a flashlight into a glass filled to the brim with water. The water does not overflow. **Energy takes up no space.**

- **Energy keeps molecules (tiny bits of matter) in continuous motion.** Fill a glass with water. Let it stand 5 to 10 minutes until the water is still. Place one drop of food coloring near the top of the water and let students observe for 5 or 6 minutes the mixing of the water with the coloring. (Lighting it from below will accent the color mixing.) The molecules of water and color strike each other due to their normal motion, and in the process they mix together. The stringers of coloring are **schlieren**.

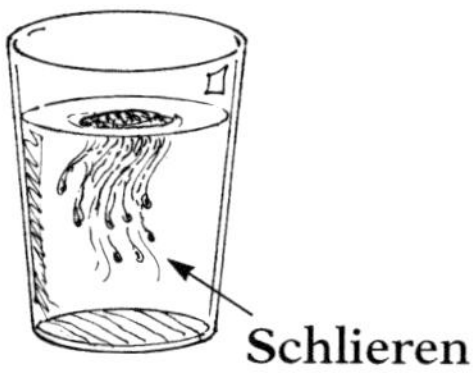

Schlieren

4. Energy (Heat) Expands Matter

When energy in the form of heat is added to solids, liquids, and gases, matter expands.

Materials:

For solids—brass ring with a tight-fitting brass ball (obtainable from scientific supply house, or make your own apparatus—see page 7), Bunsen burner

For liquids—small bottle, cork to fit bottle, soda straw, food coloring, water, florist's clay, pan with water, stove or electric coffeepot

For gases—small bottle, balloon, pan with water, stove or electric coffeepot

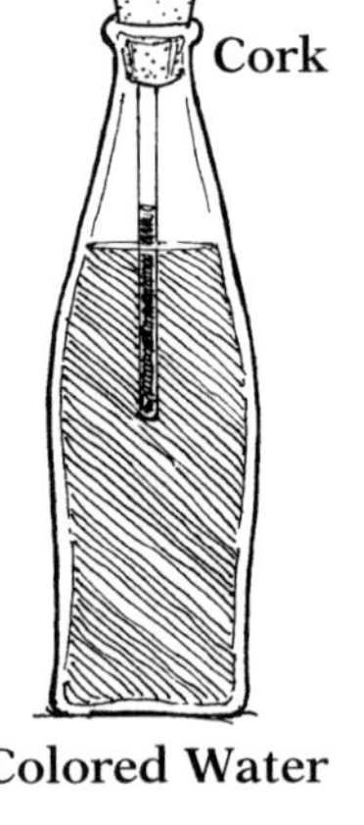

- **Solids**: Show students how the brass ball moves freely through the brass ring at room temperature. Heat the ball until it does not go through the ring. The brass has expanded. Cool the brass ball by dipping it in water. Show how it goes through the ring again.

- **Liquids**: Fill a bottle about 3/4 full with colored water. Make a hole in the cork and insert the soda straw until the straw is about 1 inch above cork. Place the cork in the bottle and seal the soda straw with clay. Notice the height of the water in the soda straw. Place the bottle in the pan full of hot water or in a coffeepot. Notice how the water level in the straw rises. The water has expanded.

- **Gases**: Place a balloon on top of a small bottle. Place the bottle in hot water. Notice how the balloon inflates. The air inside the bottle and balloon has expanded due to the heat energy. Place the bottle in cold water and notice how the balloon contracts.

Homemade Apparatus for Demonstrating Expansion of Solids

Materials: two dowels, one screw, and one screw eye

- Screw the screw 3/4 of the way into the end of one dowel. Do the same with the screw eye and the other dowel. Use these dowels in place of the brass ring and ball in the demonstration for the expansion of solids.

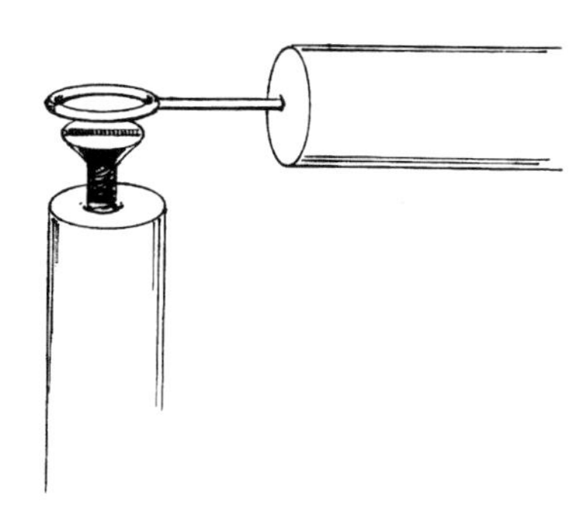

5. Absorption of Heat

Heat energy, nonvisible **infrared rays** from the sun or from other luminous sources, travels and behaves similarly to visible light. Shiny and smooth surfaces reflect heat, dark and rough ones absorb it, and transparent ones let it go right through. In sunshine, chromed car bumpers are cooler than black ones. People in the tropics wear light-colored clothing, while Arctic explorers wear dark clothes. Houses in warm climates are white to reflect heat. Low-power infrared remotes are used by the home consumer industry for controlling TVs, VCRs, etc.

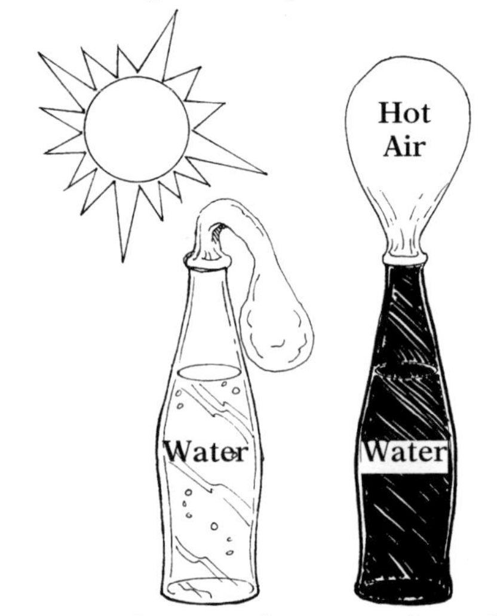

Materials: two identical bottles with a narrow neck, black tempera paint, two balloons, two same-sized empty tin cans, cardboard, water, two thermometers

- Fill the two bottles with water to the same level. Paint one bottle black. Stretch the balloons, to check for their expansion. Snap a balloon on each bottle. Let the bottles stay in the sun for a while. The painted bottle will inflate the balloon more; its air has expanded more because the black paint absorbs more heat energy. The shiny glass has passed through or reflected the infrared rays.

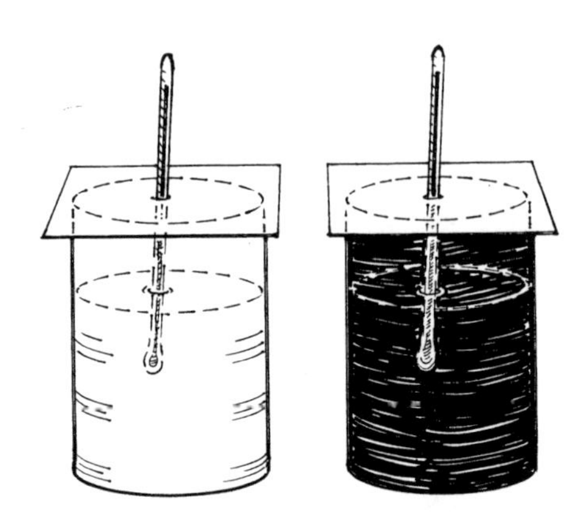

- Paint one tin can with black. Place the same amount of water in both. Cover both tin cans with a piece of cardboard. Pierce a hole in each piece and insert the thermometers. Place both cans in the sun for 15 to 20 minutes. Again, the black can will show a higher water temperature than the shiny one.

6. Radiant Energy

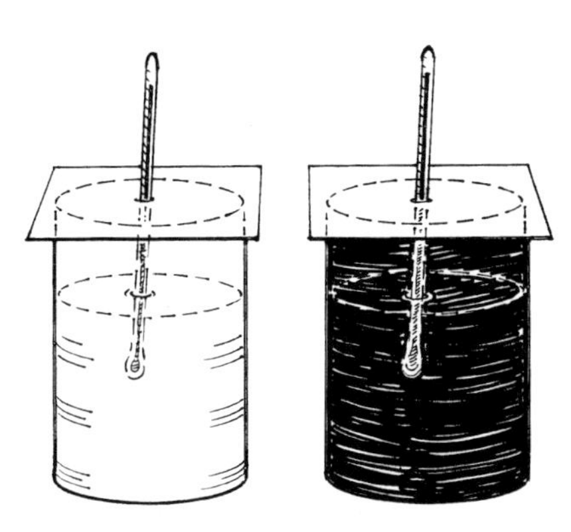

Heat **radiates** out of a substance toward a region of less heat. Heat energy goes through a transparent substance, is reflected by a smooth and shiny surface, and is absorbed by a rough or dark substance.

Materials: two identical shiny tin cans, black tempera paint, hot water, two thermometers, cardboard

- Paint one tin can with black tempera paint. Fill both with the same amount of hot water. Cover both with cardboard. Pierce a hole in the cardboard and insert the thermometers. Let the cans sit in a shady place. After 15 to 20 minutes, let your students note that the black can has radiated more heat, for its thermometer shows a lower temperature.

7. Vacuum Bottles

Vacuum bottles are an application of the principles of heat reflection. The inner part of the container is a double-walled glass jar, silvered on the inside and on the outside. The space in between the glass walls is a vacuum; most of its air molecules were pumped out. Therefore, this empty space is a poor conductor of heat. The double walls and the space in between provide three barriers to heat transmission. If you place hot liquids in the bottle, the heat is reflected back. If you place cold liquids in the bottle, heat is radiated away from the bottle, keeping it from entering. In this manner, liquids can be kept at the desired temperature.

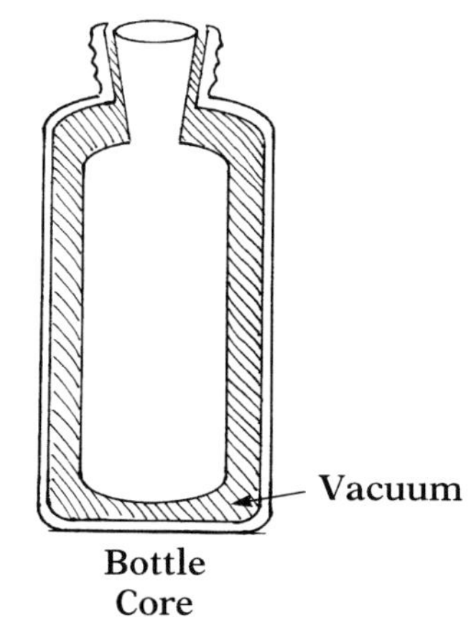

Materials: vacuum bottle, flashlight

- Bring to school an empty vacuum bottle and let your students look at it. If possible, unscrew the top and let them see the inner bottle by itself. Handle with care, for it is fragile. Shine the flashlight inside the bottle to show the silvering.

8. Kinetic Molecular Theory: States of Matter

Molecules in

Solids Liquids Gas

The kinetic molecular theory explains how matter behaves and how it changes from one state into another. Molecules in **solids** are held quite closely together by molecular bonds. All molecules vibrate. If heat energy is added, the molecules move faster and farther apart, until they begin to slide over each other. At this point, the solid has changed into a **liquid**. If more energy is added, some molecules escape from the surface of this loose bond in the liquid and begin to move even farther apart. The new state of these escaping molecules is **gas**. If energy is removed, gas changes into a liquid and a liquid changes into a solid. These are the phases (states) of matter.

Materials: ice cubes, ice cube tray, water, freezer

- Take several ice cubes and let them melt in a glass until they change into water. Ice cubes melt, for heat energy **is added**.
- Fill an ice cube tray with water and place in a freezer. Water, a liquid, becomes a solid. Heat energy **has been removed**. Changes of states are reversible.

	$+\Delta \rightarrow$		$+\Delta \rightarrow$	
Solid	**→**	**Liquid**	**→**	**Gas**
	$-\Delta \leftarrow$		$-\Delta \leftarrow$	

Δ = Energy

Materials: three identical boxes 10 inches long, clear plastic wrap, masking tape, 15 marbles

- Remove the tops of the boxes and modify two of them to be 5 inches and 7 inches long. The boxes will represent molecules in solid, liquid, and gaseous states. Place 5 marbles in each and cover them with transparent plastic wrap. Tape the wrap. Have students shake the 5-inch box. Then have them shake the 7-inch box a bit harder and the 10-inch box even harder. The behavior of the marbles resembles the behavior of molecules in matter.

- **MATERIALS:** coffeepot, water
- Take a coffeepot, fill it partially with water, and let it warm. Make certain that its lid is in place. After the water has warmed up, (in electric units a warning light goes on or off), lift the lid and let students observe how water is forming drops on the top of the lid. Have them observe the whitish cloud coming up from the pot. They will call it steam, but it is not. Steam is invisible. What they are seeing is the water vapor that forms as the invisible steam cools down and condenses into many tiny water droplets that look white. The lid, a little bit cooler than the air inside the coffeepot, also condenses gaseous water and forms water droplets. This is why water drips off the lid when you uncover the coffeepot or any other cooking pot.

MATERIALS: a glass or two, several ice cubes, insulated container

- Place a glass in an insulated container and place several ice cubes around the glass. Let the glass become chilled. Take the glass out and show the class how it frosts. This frost is the **condensation** of water in the air. The molecules of water condense since the glass is colder than the environment.

9. PRESSURE OF AIR

When energy is **unbalanced**—i.e., higher versus lower pressure, be it osmotic, electrical (voltage), or physical (air pressure)—**energy flows from the region of higher pressure to the lower one** until a balance (**equilibrium**) is reached. If a space capsule loses its pressure integrity, life-supporting air escapes out to the vacuum of space with fatal consequences for the spacefarers. On earth, the pressure of the ocean of air (atmosphere) above it is 14.7 lb/in^2. This pressure is sufficient to crush any container with a vacuum inside it, unless it is specifically designed to withstand these forces, as are Magdeburg hemispheres.

MATERIALS: hard-boiled egg (peeled), birthday candle, matches, large bottle or jar with a mouth barely smaller than the egg

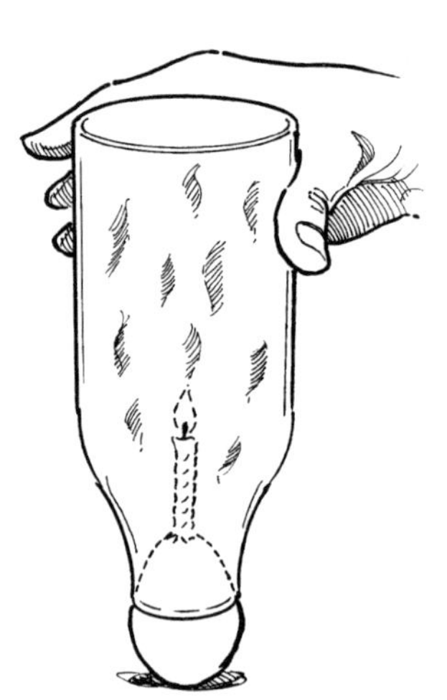

- Stick the candle into the egg on one of its ends and light it. Place the candle inside the bottle and seal the opening with the egg. As the candle burns the oxygen, 20% of the inside gas, the bottle will have a region of lower pressure. Eventually the egg will squeeze into the bottle. (During this demonstration, be careful to position your hands so that students can see how things happen.)

- To remove the egg, point the bottle opening down and blow air into it. The air above the egg will have more pressure and the egg will squeeze out.

10. Air Pressure and Soda Can

Materials: empty soda can, tablespoon, small pan with water, forceps, Bunsen burner or hot plate

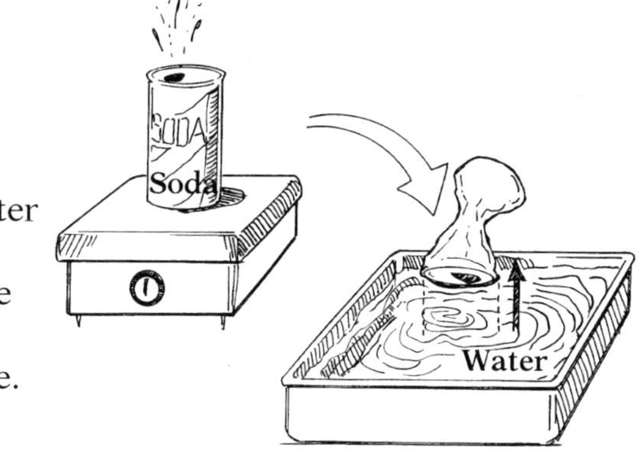

- Place a tablespoon of water into an empty soda can. Heat it on a Bunsen burner until steam begins to escape from the pop-top. Rapidly invert the can and place it in the water bath. The cooling of the air inside the can will create enough difference in air pressure that the can will be crushed by the atmospheric pressure. This happens because the steam filling the can contains fewer molecules than the equivalent volume of hot air. Thus, when the gas inside the can is rapidly cooled, it creates a vacuum.

11. Work from Air Pressure

The pressure of air can be made to do useful work. One example is the aerosol bottles used for perfumes, deodorants, paints, etc. Divers use tanks of compressed air. There are many compressed-air (pneumatic) tools in many shops: air hammers, air buffers, air ratchets, air drills, etc. Compressed air is used in tires, footballs, basketballs, air brakes, riveters, and sandblasters.

Materials: bottle, cork with hole, eyedropper, small piece of rubber hose to fit dropper, water

- Insert the dropper in the cork so that the narrow part is up, leaving a small portion extending from the bottom side. Connect the rubber hose to the bottom of the dropper. Cut the hose just long enough to reach to the bottom of the bottle. Half-fill bottle with water and close the bottle with the cork. Blow vigorously into the bottle. Within seconds, the bottle will become a fountain. The air fills the space above the water and becomes compressed. The pressure pushes the water out. This is similar to how aerosol bottles work.

12. Light Travels in a Straight Line

Light travels in a straight line. Our sun is the original source of all energy on our planet, including light.

Materials: several index cards, small pieces of clay, flashlight, hole punch, soda straw

- Punch a small hole in the center of the cards. (Draw diagonal lines to find the center.) Stand all cards vertically by pushing them into a blob of clay. Place the flashlight on one end and arrange the cards so that light can be seen through all the holes. The holes must be arranged in a straight line if light is to go through them. Experiment with finer holes.

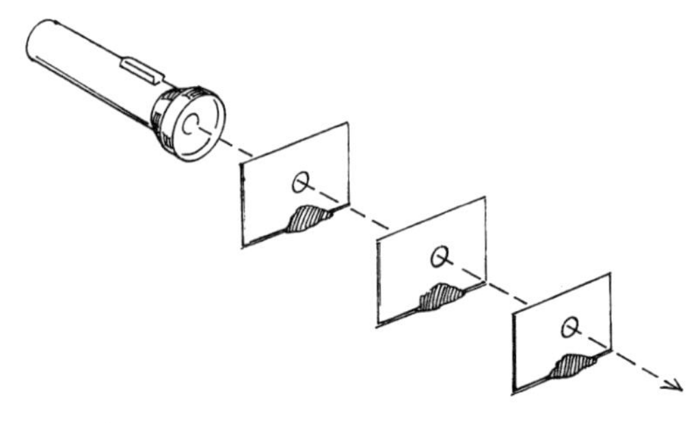

- Have your students look through a soda straw at the flashlight in a darkened room. Have them bend the straw, and no light will go through it.

13. Pinhole Camera

A pinhole camera is another way to demonstrate that light travels in a straight line.

Materials: one small cardboard box, a small piece of aluminum foil, wax or tissue paper, adhesive tape, a needle, flat black paint

- Take the following steps:

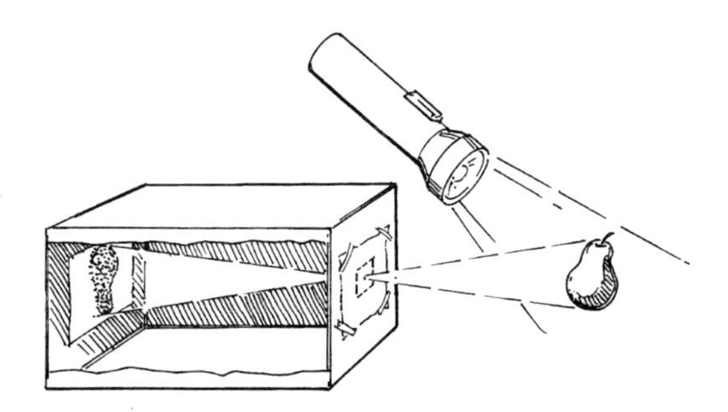

1. Cut a hole about 1 inch square in one end of the box. Make certain that it is in the center. (Drawing diagonal lines will help locate the center.)
2. In the opposite end of the box, cut an opening at least 4 inches square. Try to have it centered.
3. Cover the smaller hole with flat aluminum foil, and tape the edges of the foil flat to the box with the adhesive tape.
4. Paint the inside of the box with flat black paint and let it dry.
5. With the needle, pierce a small hole in the center of the aluminum foil.
6. Cover the larger hole with tissue or wax paper.
7. Place a small object in front of the pinhole camera and light it with the flashlight. Make certain that the flashlight is above the camera.
8. You will be able to observe the object on the paper in the back of the pinhole camera. The image will be inverted. This is additional proof that light travels in a straight line.

- Move the camera closer and farther away from the object. Show your students what happens to the size of the image in the back of the pinhole camera.

14. ANGLE OF INCIDENCE EQUALS ANGLE OF REFLECTION

A key property of light is that when it strikes a reflecting or shiny surface, its angle of **incidence** (strike) is the same as its angle of **reflection**. A **normal line** is a line perpendicular (at 90°) to the reflecting surface. If the surface is smooth, like a mirror or a very calm body of water, there is a reflected **regular image**. If the reflecting surface is rough or the water is moving, then the **image** is **diffused** and no image can be seen. When photographing people or objects in front of shiny surfaces like mirrors or glass, photographers take the shot at an angle to the surface, to avoid their own reflection and that of their flashes. Photographers also use very high flash brackets to avoid having people with red eyes in photographs. The flash has a high angle of incidence and its reflection is at a low angle, so that the camera does not see the flash's reflection in people's retinas. In photography, the minimum successful working height of the flash above the lens is 13 inches.

MATERIALS: flashlight, black paper, mirror, clay, masking tape, razor blade, protractor, small ball

- Before you do the next demonstration, bounce a ball on the floor at an angle. It will bounce back at a similar angle. Discuss with your students that people who play pool and billiards are very careful about the angle at which they hit the balls. If the billiard balls hit the table's side at an angle they will bounce back at the same angle, unless they have some spin (English) applied to them.

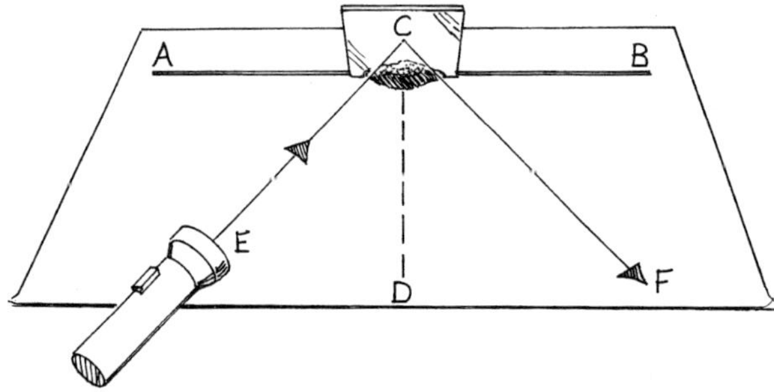

- Cover the flashlight with black paper and cut a narrow slit in the paper. Place the mirror vertically by sticking it into a blob of clay. On a larger piece of paper, draw a long straight line *AB*. Somewhere in the middle, mark a point *C* and draw a line at 90° to *C*. Label it *CD*. This is a normal line to line *AB* at point *C*. Place the mirror's **back surface** parallel and on top of line *AB*, centered on point *C*. Keeping the slit vertical, shine the flashlight on its narrow beam at point *C*. Mark both the ray of incidence (striking) light and the ray of reflected light. With a protractor, measure the angles of incidence and reflection. They should be the same. (If they aren't, the mirror was not vertical.)

15. Mirror Images: Depth

An **image** in a mirror appears **reversed**. The **real distance** of an object in front of a mirror is its distance from the mirror. **Depth distance** is the distance of a virtual image "behind" the mirror. The two distances are identical. If you want to photograph a person in a mirror, you have to adjust the camera focus for twice the distance between that person and the mirror. In flash photography, this depth distance becomes critical. If you are 10 feet from a mirror and so is your subject, you have to set the flash for 20 feet.

Materials: small mirror, two books, pencil, ruler, masking tape, construction paper about 20 × 20 inches

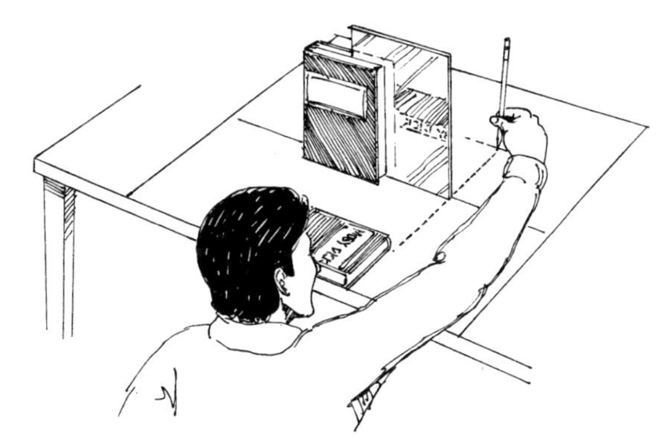

- Place a book or some written material in front of the mirror. In the mirror, it appears in reverse.

- Take a piece of construction paper about 20 × 20 inches, and place it on a table with one edge at the table's edge. Tape it in place. Draw a line across the middle of the paper, parallel to the table's edge. This is your reference line. Place the mirror in a vertical position by wedging it in a book. Place the mirror assembly on top of the reference line. Make certain that the back edge of the mirror is on the reference line. Place a book in front of the mirror with its written edge facing the mirror, about 6 inches away. Place your eye near the edge of the table. Look with both eyes at the mirror and the image of the book's edge in it. Move a vertical pencil beside the mirror, on one side. Move the pencil back and forth until it lines up with the book's reflection in the mirror. Mark this spot with the pencil. Measure the distance between the pencil mark and the rear edge of the mirror. Measure also the distance of the book's edge to the mirror. The two measurements should be nearly identical. (If the mirror is not vertical, the measurements may differ.) Use the back edge of the mirror as a reference for all your measurements.

16. Bending Light: Laws of Refraction

Rays of light change their direction when they go from one transparent medium to another. This change of direction is **refraction**.

1. If light moves at an angle from a less dense medium into a denser one (for instance, from air into glass), it bends toward the normal line.

2. If light moves at an angle from a denser medium to a less dense medium (from water to air), it bends toward the normal line.

3. If light moves along the normal line, like a beam hitting a windowpane on the perpendicular, it is not bent.

Materials: empty coffee can or dish of similar size, coin, water, fish tank, sheet of glass to cover part of fish tank, flashlight, black paper, razor blade, masking tape

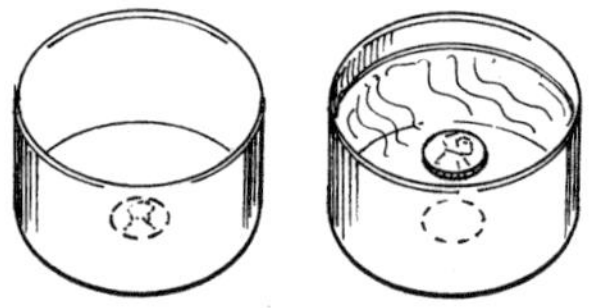

- Place the coin on the bottom of the coffee can. Move slightly away until the coin just disappears from sight, blocked by the edge of the can. Have a student pour some water in the can as you stay in the same place, and watch the coin's image reappear. As the reflected light of the coin passes from the water to the air, the rays of light bend away from the normal toward your eyes.

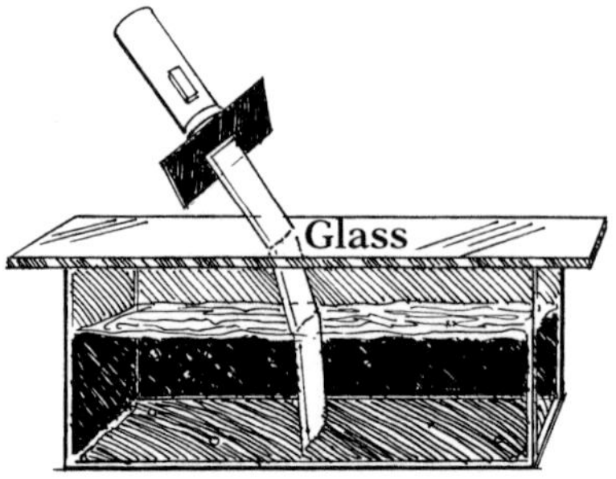

- Add water to the fish tank up to about 3/4 full. Cover one end of the tank with a sheet of glass. Prepare the flashlight by covering it with black paper and cutting a narrow slit in the black paper. In a darkened room, shine the flashlight's narrow beam at an angle through the glass into the water. It will show refraction.

17. MOTION PICTURE EFFECT

Motion pictures are possible due to a defect of the human eye. The eye will continue to see an image for 1/15 second after objects have been removed from view. This is **persistence of vision**. By projecting still images rapidly, so that each new image comes to view prior to the extinction of the previous one, the eye sees images "moving." On television, the images are formed by scan lines, 525 per picture. The picture tube draws the odd and even lines alternately, 60 times a second, so that the persistence of vision makes it appear as a completed moving picture.

MATERIALS: pencil, razor blade, pin, index card

- Slit the pencil eraser in the middle carefully. Draw an image on each side of the index card. Insert the card into the slit and fasten with the pin. Twist the pencil rapidly with your hands, and both images will appear as one.

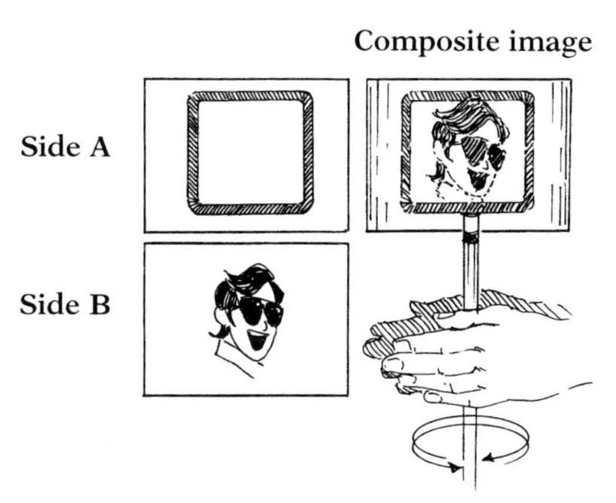

18. Electromagnetic Spectrum

Visible light is a small part of a larger spectrum of energy: the **electromagnetic spectrum**. **Electromagnetic waves** are transverse waves, which vibrate perpendicular to the direction of travel and consist of oscillating (vibrating) magnetic and electric fields. They have a wide range of frequencies and can travel through all media, including a vacuum. When they are absorbed, they cause a rise in temperature. With the exception of radio waves, these energies occur in random pulses (called **photons**), not in a continuous stream; they are explained by **quantum theory**.

Table: Electromagnetic Energy

Cosmic rays	Background radiation; **particles** of enormous energy given off by stars.

Gamma radiation	Deadly high energy given out by the sun and other stars
X rays	High energy used in X rays
Ultraviolet rays	Invisible energy waves in sunlight, which cause skin to tan
Visible light	Basic colors of light, emitted by the sun and visible to the human eye
Infrared rays	Rays of heat energy; sensed by our nervous system
Radio waves	Microwaves; TV; radio energy

Materials: small lamp with bulb

- Show students a lightbulb in a lamp. They will immediately see the visible light, while they will have to feel the heat—an invisible form of energy. Other forms of energy need special detectors.

19. Frequency, Wavelength, and Amplitude

Frequency (number of waves per second) varies inversely with **wavelength**, the width of a wave. As the number of waves (frequency) increases per unit of time, the wavelength becomes smaller. (See the diagram that follows. This is a key diagram to show your students so they can understand this important relationship.) Frequency is measured in hertz (Hz) waves per second in honor of Heinrich Hertz, who did pioneering work in this field.

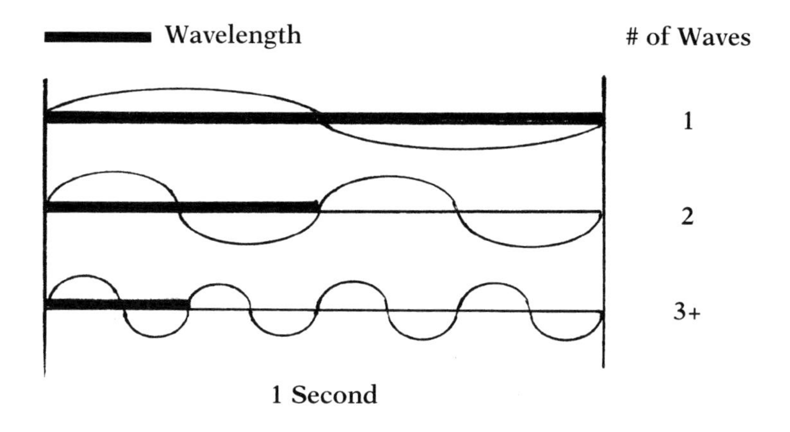

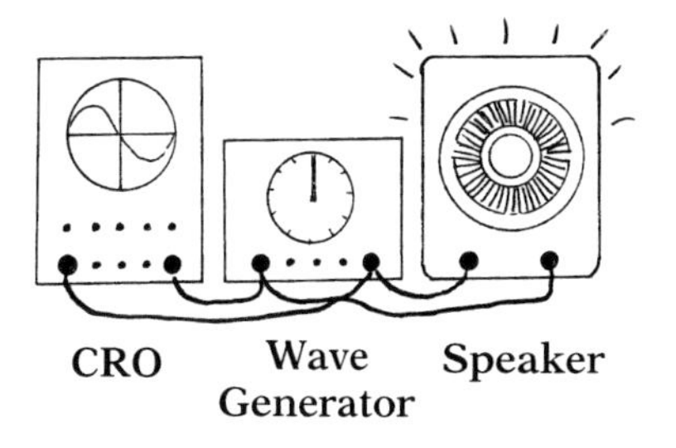

Materials: cathode ray oscilloscope (CRO), sine-square wave generator, speaker, four to six wires with alligator clip ends

- Hook up the output of the wave generator to the input of the CRO. Connect the same output points to a separate speaker. Select low frequencies in the audio spectrum so students can hear the actual sounds through the speaker. Show a few, clean sine waves on the scope. Increase the frequency of waves and notice how the width of the individual waves becomes narrower to fit in the display. The waves have a shorter wavelength.

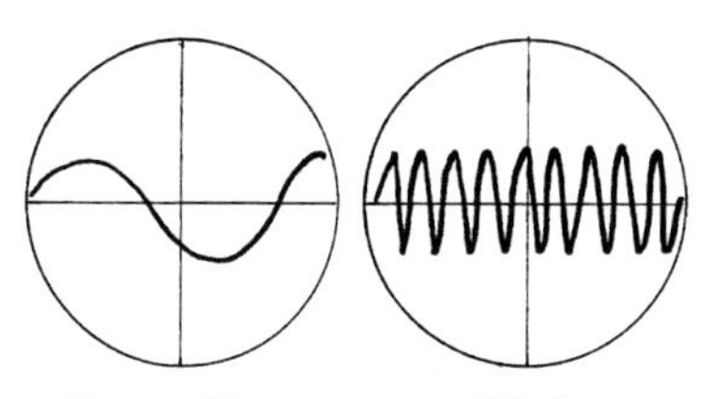

- Vary the output on the sine generator by changing its volume control. The sound will go from barely audible to loud. Observe the change in height of waves. The loud waves are higher than the soft ones. The **height** of the wave above or below the reference line is its **amplitude**. Amplitude measures the amount of force in the wave. A tall ocean wave has more power than a small one.

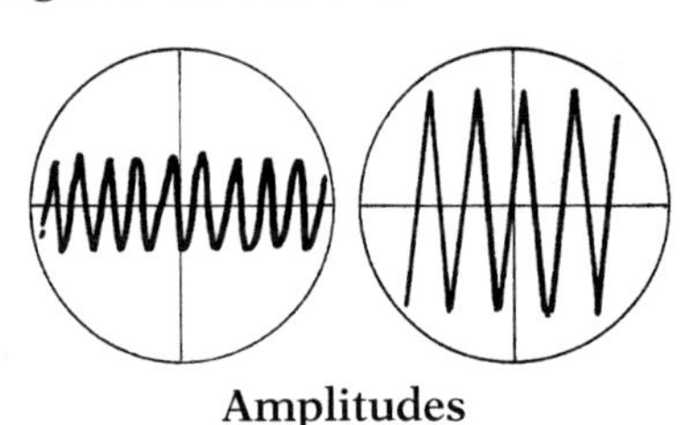

Amplitudes

20. THE LIGHT SPECTRUM: COLOR OF OBJECTS

While most components of the electromagnetic spectrum are invisible, **visible light** can be seen. Light is a mixture of colors, and its color depends on its wavelength. Violet has the shortest wavelength, while red has the longest one. The other colors are in between these two. Ultraviolet light has shorter waves than violet light and is invisible. It causes skin to tan and it makes fluorescent objects glow. A longer wavelength than red is infrared. It is invisible heat energy, and one can only feel it. **Objects absorb all colors and reflect their own.** Black absorbs all colors, white reflects all colors, and red reflects only red. Transparent and translucent objects transmit their own color and block the rest. They act as color filters. If you shine a white light through a red filter, it will let only red light pass.

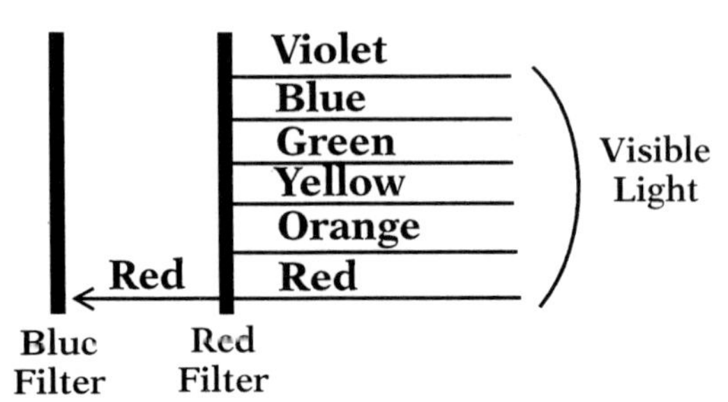

MATERIALS: several color filters, colored pictures, and pictures from magazines

- Look at the various colored pictures through the color filters. You will see only those components of the pictures that have the same color as the filter; the rest will be either white or invisible.

21. LASER LIGHT

Laser light is different from ordinary light.

TABLE: PROPERTIES OF LIGHT

ORDINARY LIGHT	LASER LIGHT
Small amount of energy	Huge amount of energy
Many different wavelengths	Single wavelength
Scatters out	Does not scatter out (except over interplanetary distances, and then slightly)
Many colors mixed	Single color
Cool	Hot, melts metals
Hard to focus in narrow beam	Narrow and extremely fine beam

Due to their unique properties, lasers are used in surgery, industry, consumer products, and space communications. Nowadays, grocery stores have laser scanners at checkout counters for reading bar codes. Music and video are recorded on disks that read music and video with laser light. The police use laser guns to check traffic speeds. Laser light is used to make 3-D (dimensional) photographs called holograms. Holograms are used on credit cards, postage stamps, driver's licenses, and many other applications.

MATERIALS: regular two-cell flashlight, small narrow-beam flashlight, hologram stamp (if available from post office)

- Shine the large flashlight on the ceiling of your darkened classroom, and move the beam away from you. As it moves, it will spread out considerably. This illustrates regular (incoherent) light. Shine the small narrow-beam flashlight on a spot above you and hold it steady. Explain that a laser beam would be only a tiny dot, as small as a pin. (For reasons of safety, it is best that you do not use a real laser in the classroom, despite its low cost.)

22. Lenses

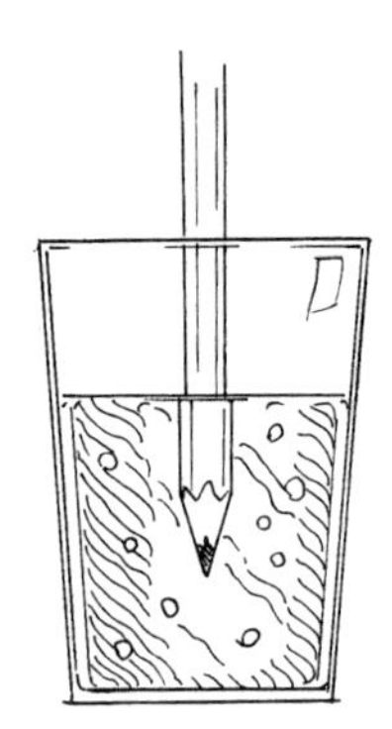

Lenses transmit light. In the process, they slow down light rays. Some lenses transmit light as is, while others make objects appear either enlarged or reduced. Microscopes are used to enlarge small objects. Lenses scatter light and make objects look less bright than they really are.

Materials: a tumbler, some water, and a pencil or a key

- Build a simple microscope (one lens) by filling a glass with water and dipping an object, such as a pencil or a key, into it. Note how the object is enlarged. (Bottom lighting will improve the effect.)

- An alternate demonstration: Give students a paper clip. Have them bend it to form a small loop on one end. Have them fill the loop with a small drop of water. Use it as a magnifier to look at their hands or anything around them.

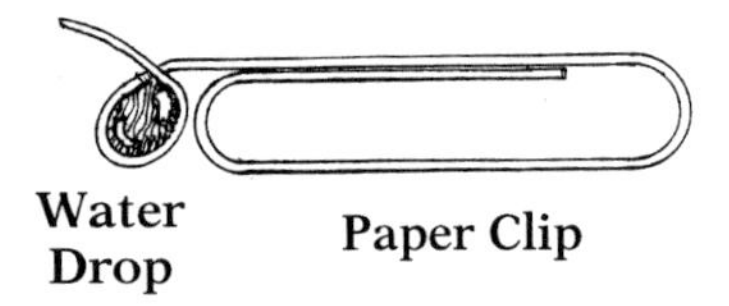

23. Refraction of Light #1

When light goes from one medium (transparent substance) into another, light bends: **refraction**.

Materials: tumbler, pencil, water

- Fill a glass nearly full of water and place a pencil in the water at an angle. Notice how the pencil seems to bend. This bending is **refraction**.

24. Refraction of Light #2

Materials: beaker, water, coin

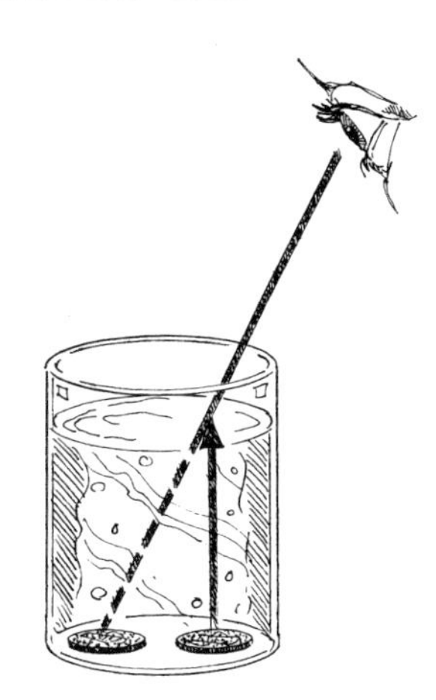

- Place a coin at the bottom of a beaker nearly full of water. Holding the beaker slightly to the side, look down at the coin. The coin will appear where it is not, and you will see two coins. Shake the glass lightly to see which coin does not move. Notice that the real coin appears larger, since the water acts as a lens and magnifies the image.

25. Breaking Light Apart

The light we see combines all the colors put forth by the light's source. To break the light into its individual component colors, you use a **prism**, a triangular piece of glass or plastic, or a **diffraction grating**. A diffraction grating is a piece of plastic or glass with thousands of fine lines scratched on its surface. Rainbows demonstrate this process in nature; drops of rainwater, acting like prisms, break the sunlight into its component colors.

Materials: Two pieces of cardboard (at least one of them white), razor blade, masking tape, two prisms (or one prism and a magnifying glass), film projector, black paper, clay

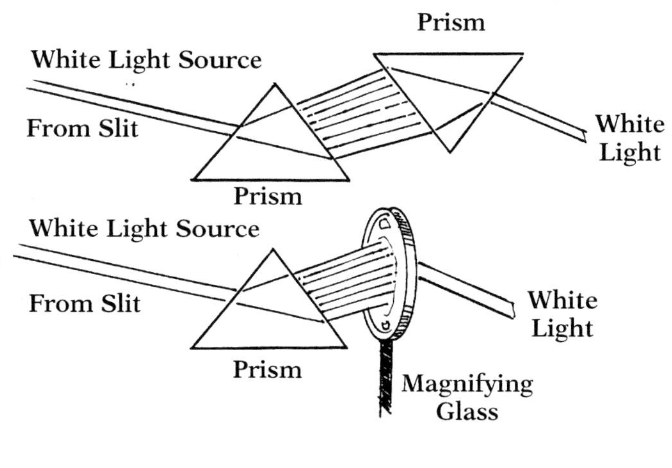

- The following activities are done either with sunshine or a projector. If sunshine enters your classroom, do part A. If sunshine does not enter your classroom, then do part B. If you do not have a prism, do part C and use either light source.

A. (1) Tape onto your window a piece of cardboard with a slit about 1 inch high and 1/8 inch wide. Place a table nearby so that the rays of the sun, going through the slit, hit the table. Make a base for the prism out of the clay; place them both on the table. Turn the clay until the prism projects a band of rainbow colors onto the piece of white cardboard, which you hold nearby.

(2) Place the magnifying glass or the second prism in the path of the rainbow colors and project them through to the white cardboard. This should recombine the colors together into the original light.

B. Repeat all procedures as in part A, except use a projector as your light source. Cover the lens with the black paper and cut in it a vertical slit, as fine as you can.

MATERIALS: water, small pan, mirror

C. Fill the pan with water and place the mirror under the water, leaning on one side of the pan. Let the light come through the slit and strike the mirror. The mirror will project the spectrum on the wall below the slit. It may take a little effort to aim the mirror.

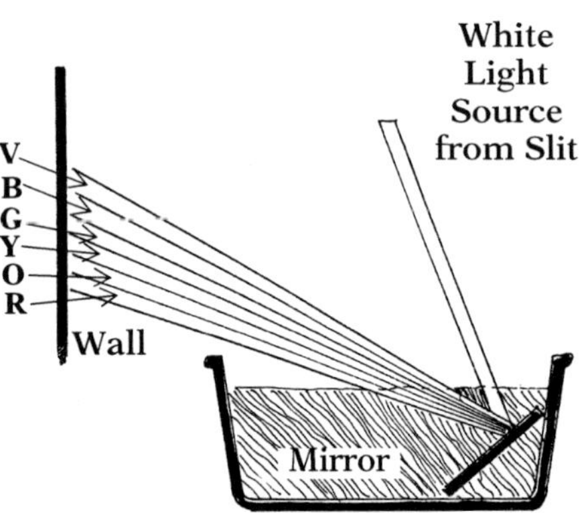

26. POTENTIAL ENERGY #1

Potential energy simply means energy at rest, ready to become motion energy (**kinetic energy**) to do some work.

MATERIALS: two books, marble

- Place a marble on a ramp made by a pair of books. While you hold the marble near the top of the book, the energy of the marble is potential. It has energy because of its position (height) above the bottom of the book. When you let go of the marble, it rolls down and away, for the **potential energy** has changed into motion energy: **kinetic energy**. Objects that are raised from the ground have potential energy. When they fall they have kinetic energy. The fall is due to the pull of gravity, thus they have **potential gravitational energy**.

27. Potential Energy #2

Potential energy is increased if the height of the marble in the previous demonstration (26) is increased by making the ramp steeper. **Height** is the key variable. Work is done to bring the marble up the ramp. The higher the marble, the more work it took to get it there; thus, the more the stored energy it has.

Materials: empty plastic gallon bottle, pencil-type soldering iron, water, sink or suitable container to catch water

- Take an empty plastic gallon container and make about six to eight equally spaced holes from the bottom to the top. (A pencil soldering iron will melt circular holes.) Fill gallon with water rapidly, and observe how the water spurts out of the holes. As the height of the water column increases, so does the pressure and thus the spurts get longer. The height of the water column creates the **potential energy**; the length of the spurts shows the kinetic energy.

28. Speed, Velocity, and Friction

Speed is distance traveled per unit of time (e.g., per second). If the speed is in a specified direction, then it is called **velocity**. To find the average speed, divide the distance traveled by the time taken to travel it.

Materials: Frisbee™, measuring tape or meterstick, calculator, marble, several books, masking tape, stopwatch

- Throw a Frisbee from a reference line and measure how long it takes to land. Measure its distance from the starting point. Calculate the speed by dividing the distance by the time in seconds.

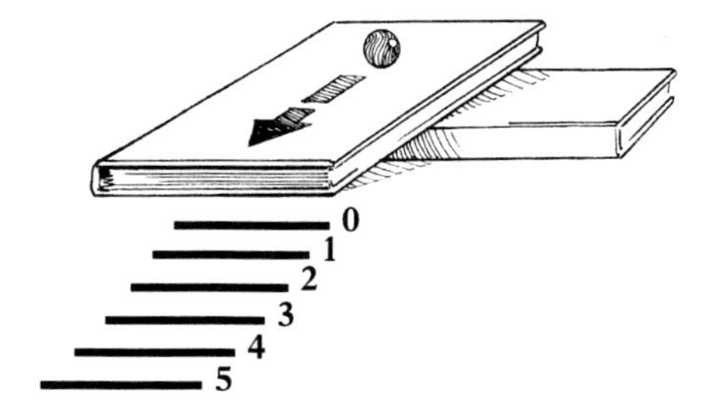

- Make a ramp on a flat surface with a book, and mark distances in 1 m increments from it. Launch a marble down the ramp and time it as it rolls. Divide the distance by the time to obtain the speed. For ease of calculations, measure the time from 0 to 1 m, 1 to 2 m, 3 to 4 m or any other choice. Be aware that the marble will slow down due to **friction**, a force opposing motion. Note that the average speed of the marble will decrease with its distance from the starting line.

- To demonstrate friction, redo the same activity on a rougher surface, like carpeting or concrete, and compare the results. Friction will slow down everything.

29. Acceleration

Acceleration is the change in speed per unit of time. (Note that acceleration can refer to something slowing down; this is known technically as negative acceleration, but is generally called deceleration.) To calculate acceleration, subtract the original velocity from the final velocity and divide the result by the time. One unit of velocity is m/sec. One unit for acceleration is m/sec/sec or m/ sec^2. The second power is your clue that the unit is acceleration. Acceleration is a gauge of how rapidly speed changes. Racing cars can go from 0 to 60 mph in a few seconds, while most cars take longer. The difference is the racers' ability to accelerate, due to their more powerful engines.

Materials: books, marbles, masking tape, metric ruler, stopwatch

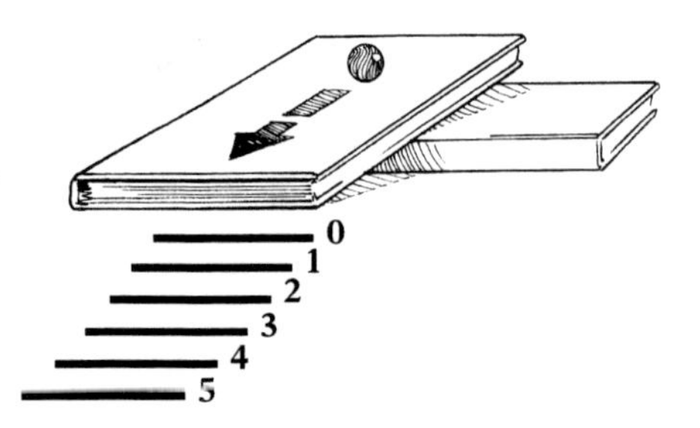

- This demonstration will take a full period. Set up a ramp system like the one shown in the illustration. Make certain that you have at least 5 m of run. Test the ramp to make sure your marble will roll at least 5 m. I use a 5 m counter. The floor is fine if it is smooth. Otherwise try a spot outside your classroom. The entire class participates with you in this demonstration. Some students are the timekeepers, some are launchers, and some collect the marbles, while everyone else records the data and calculates the results. The goal is to find the average speed for each 1 m interval. Repeat all measurement three times and then average them. The difference in the average speeds between the intervals is the average acceleration.

Copy and complete the following data:

	TIME	**TIME**	**TIME**	**TIME**	**TIME**
	0–1 m	1–2 m	2–3 m	3–4 m	4–5 m
Launch 1					
Launch 2					
Launch 3					
Average Time					
Speed	A	B	C	D	E
Acceleration	A–0	A–B	B–C	C–D	D–E

The capital letters in the boxes represent the calculated speeds. To find the accelerations, subtract the speeds as indicated. The final unit will be m/sec/sec or m/sec^2. Note that the original speed is 0.

30. BERNOULLI'S PRINCIPLE

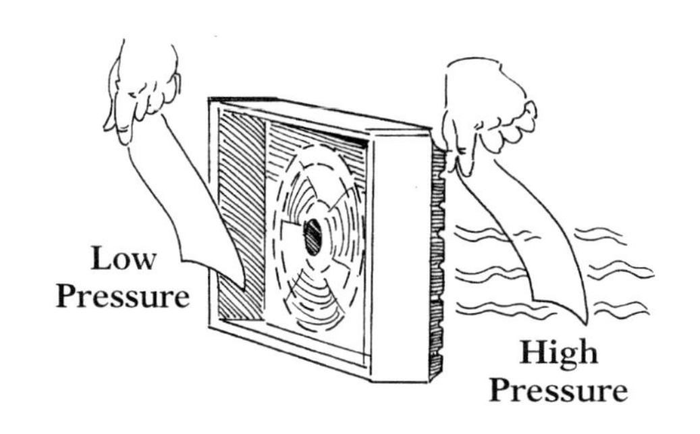

Bernoulli's principle states that when the speed of a moving fluid **increases**, the pressure on its edges **decreases**. This principle explains why planes, birds, and kites fly. It explains many things we see in our lives, from the action of rocket motors to the action of paint spray guns.

MATERIALS: electric fan, a strip of paper, Ping-Pong™ ball, funnel, small box full of polystyrene packaging particles, vacuum cleaner hose 5 to 8 feet long, hard-boiled egg, glass, deck of playing cards

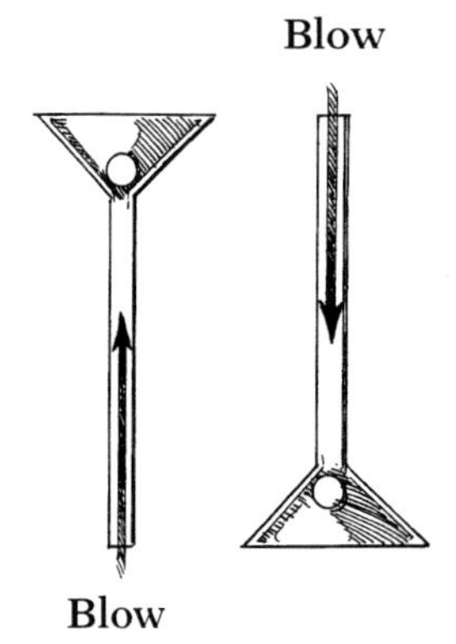

- First, demonstrate the effects of high and low air pressure: Turn on the fan and show the paper strip in front and behind the fan blades.
- Put the Ping-Pong ball in the top of the funnel, and blow upward into the stem of the funnel. The ball will not blow away but will cling to the

funnel. Invert the funnel so that it points downward. Hold the ball temporarily. As you start blowing hard, slowly remove your hand. Again, the ball is going to cling to the funnel. As the air moves away from you faster, it creates a low-pressure area in the center.

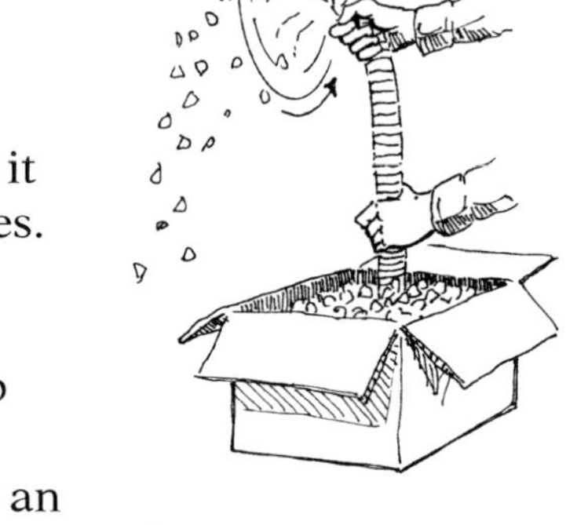

- Take a piece of vacuum hose and place it in a box containing polystyrene particles. Spin the opposite end of the hose in a circular motion. As you increase the speed of the end, the pressure will drop and pieces of foam will come out as if sucked up by a vacuum cleaner. This is an outstanding demonstration, but it requires a cleanup.

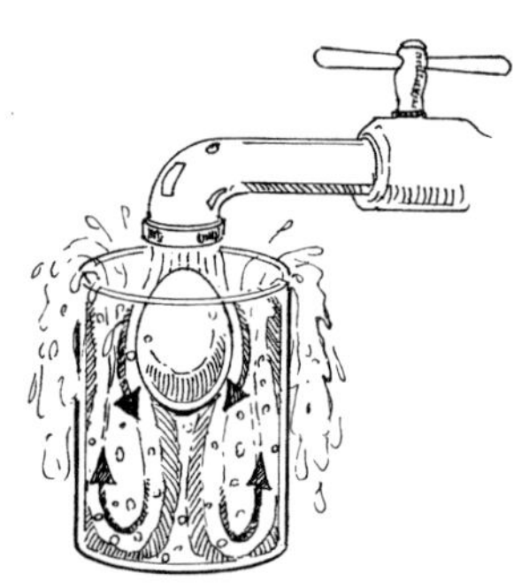

- Place an egg in a glass and move the glass under a faucet open at full power. (Remove the aerator if there is one; this activity needs a solid stream of water.) Maneuver the egg into the water stream. The egg will rise against the stream of water. The increased speed of water splitting around the egg causes lower pressure.

- Ask one of your students to hold a playing card and to make it fall straight down. The student will have trouble with this. Now, you hold a playing card horizontally. Let it go. It will come practically straight down. Hold the card on its side. It will come down flipping over and off to the side. A flat card is a wing. As it falls it compresses air below it, and above it there is less pressure. A vertical card creates a turbulent flow and does not fly.

31. Newton's Third Law of Motion

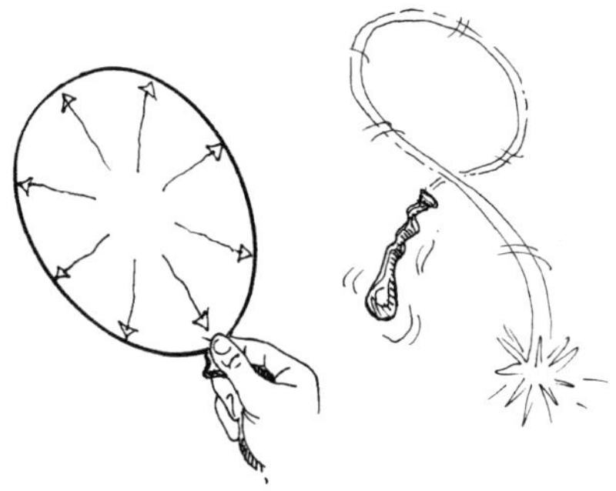

Newton's third law of motion states that for every action there is an equal and opposite reaction.

Materials: two balloons, string, soda straw, masking tape

- Inflate a balloon. Hold it pinched so that no air escapes. Inside the balloon, air exerts the same pressure on all sides of the balloon. Let go of the balloon and observe how it flies around erratically until all the air inside has gone out. As the compressed air (a region of higher pressure) escapes into a region of lower pressure, it creates a force to provide balance (equilibrium). An equal and opposite force pushes the balloon forward.

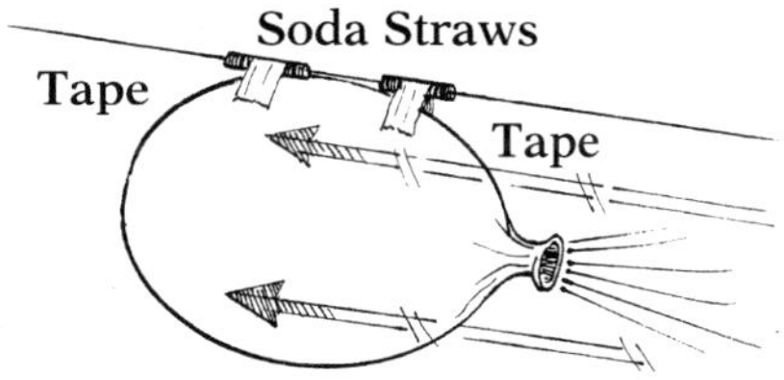

Soda Straws
Tape
Tape

- Stretch a piece of string across the classroom. Before tying the string taut, insert the string into two 2-inch soda-straw pieces. The string should be roughly horizontal, but pitching it slightly downward (in the direction of travel) will help. With the assistance of students, tape an inflated balloon to the two pieces of soda straw with the balloon's exhaust parallel to the string. Be careful not to attach the masking tape to the string. Pull the entire assembly to the end of the string and let go. The balloon should be able to cross the room. (It may take some practice.)

32. Magnets and Poles

Magnets attract only iron, cobalt, nickel, and some materials made with these elements. A bar magnet's end is a **pole**. One is the north pole and the other is the south pole. Magnetic poles follow the same rules as electric charges: Like poles repel, unlike poles attract.

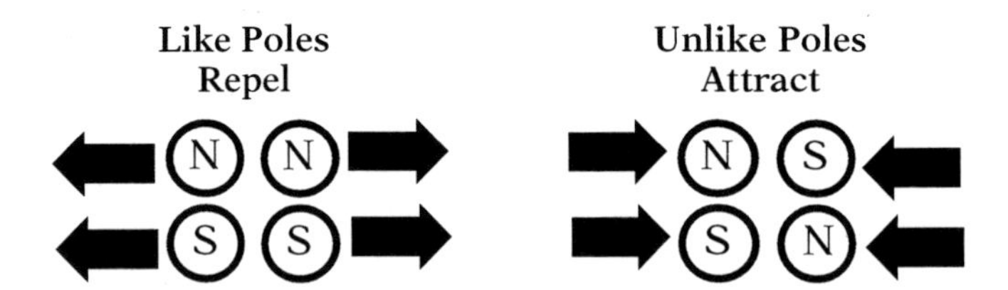

Materials: two bar magnets, paper clips, string, pencil, masking tape

- Tape the pencil to the edge of your table. Tie a magnet to the pencil so that it balances. It will slowly align itself along the north-south meridian and act as a compass. Take the other magnet and bring one end of it closer first to the north pole and then to the south pole, showing how it attracts one and repels the other pole.

- Put a handful of paper clips along a line. Place the magnet over them and lift the paper clips. More paper clips are picked up at the ends, showing that there is more magnetic force near the poles.

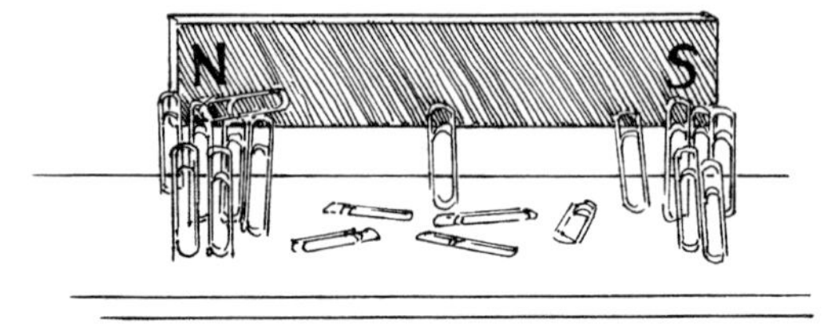

33. Making a Magnet

Atoms are like tiny magnets turned in random directions. If they are all lined up, their magnetic forces combine to make a material (iron, nickel, or cobalt) magnetic. To **magnetize** an object made of iron, nickel, or cobalt, gently pull a magnet along its length and do not remove the magnet until it reaches the end. Repeat the stroke at least 10 times. You are lining up the atoms inside the object. By stroking it several times, you are increasing the number of atoms lined up. When atoms are lined up, their magnetic forces add up. When you stroke a nail from head to point with the north pole of a magnet, the point of the nail will become a south pole. Most magnets that you prepare from iron materials will not keep their magnetism very long. Magnets made from alnico are much stronger magnets, and they keep their magnetism for years. They are permanent magnets. Alnico is made of nickel, aluminum, cobalt, copper, and iron.

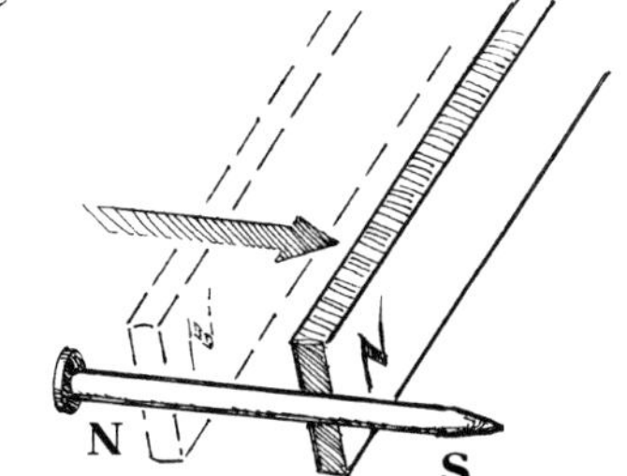

Materials: bar magnet, several nails, paper clips

- Use a nail and show that it does not lift a paper clip. Stroke the nail with the magnet 10 times. Lift one or two paper clips with the nail.

34. Inducing Magnetism

When a magnet touches a material that transmits the magnetic force, it creates **magnetic induction**. Iron and steel are such materials.

Materials: bar magnet, several paper clips

- Hang one paper clip from the magnet. Gently pick up another paper clip by using the end of the first paper clip. The magnetic attraction is stronger than the force of gravity, and the paper clips will stick to each other.

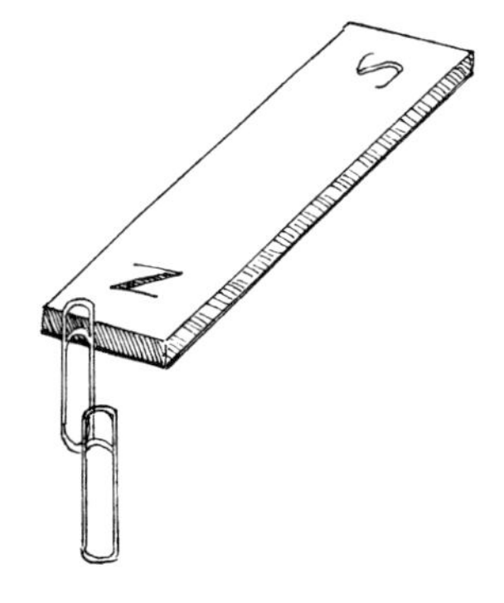

35. MAGNETIC FIELDS

Magnetic **force lines** are invisible, and they travel from one pole to the other without touching each other. All the magnetic force lines together are a **magnetic field**.

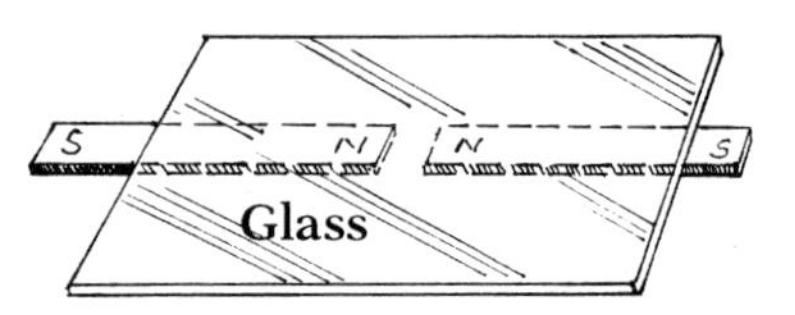

MATERIALS: two bar magnets, horseshoe magnet, iron filings, a piece of stiff acetate or a sheet of glass

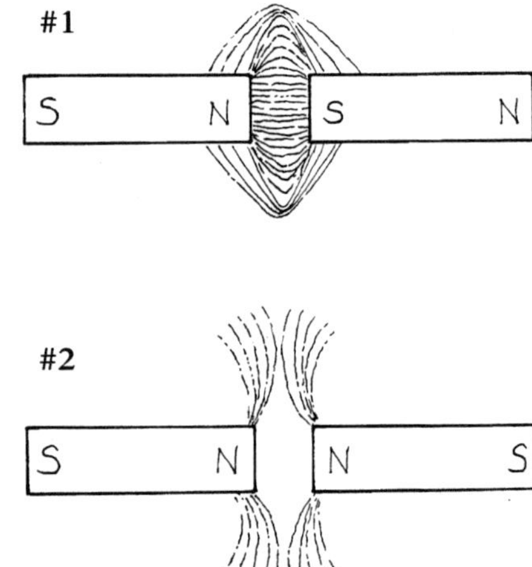

- Place the bar magnets on your overhead projector with the opposite poles about 1 inch apart. Cover them with a sheet of acetate. Sprinkle the sheet with iron filings from at least 1 foot up. Gently tap the acetate, and the lines of magnetic forces will become visible. Refocus your projector to show the magnetic lines and the magnetic field. Repeat with like poles, either N-N or S-S, or with a horseshoe magnet.

36. Magnetic Compass

Scientists consider earth to be like a giant magnet. Earth has two large magnetic poles, one near the South Pole, the other in Hudson Bay about 1000 miles away from the North Pole. The earth behaves as if it had a giant magnet running through it from the North to the South Pole. As a result of this, the planet has magnetic lines that affect all magnetic materials. Deep in the abyssal depths of the Atlantic Ocean, ferrous particles frozen in former magma near the edges of tectonic plates clearly indicate from their orientation the earth's magnetic fields and their reversals over millions of years.

Materials: pencil, string, tape, bar magnet, masking tape, sewing needle, cork, small bowl, water

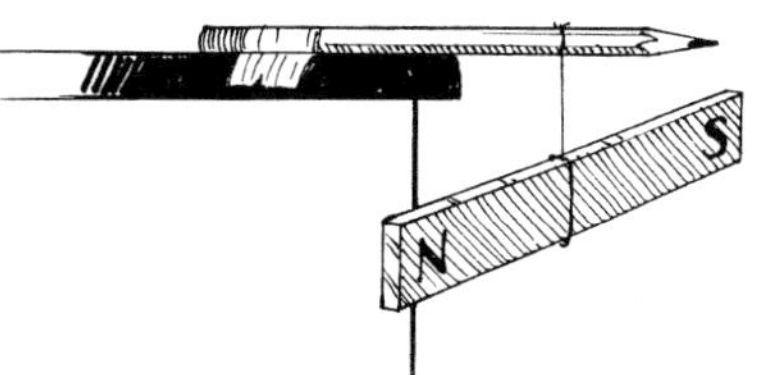

- Tape a pencil to the edge of a table, and tie a bar magnet to the pencil so that it balances. (Alternatively, use a ruler stuck between the pages of a book. Cover the book with a couple of additional books to hold things in place.) The bar magnet will line itself up along the line of earth's magnetic field.

- If you wish to use a horseshoe magnet, take a small piece of wood, about 3 inches square, and drill a hole in its middle to fit a pencil. Glue the pencil in place. Balance the horseshoe magnet on the pencil tip. Slowly the magnet will line itself up along the earth's magnetic meridian.

- Stroke the sewing needle a dozen times until it becomes a magnet. Push it through the cork and place it in a small bowl nearly full of water. You have just made a water compass.

37. Destroying a Magnet

To destroy the magnetism in a magnet, the magnet must be hit, dropped, or heated. Breaking a magnet into two only makes smaller magnets. Magnetic tapes are demagnetized (erased) by bulk demagnetizers. These magnetize the tapes at 90° to the original lines, so that the magnetic reading heads in VCRs and tape recorders do not read it.

Materials: coat hanger wire, strong magnet, compass, hammer, wire cutters, tongs, Bunsen burner

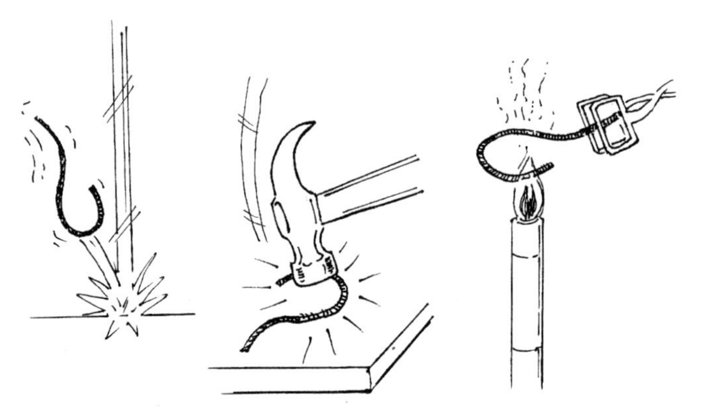

- Cut a piece of coat hanger wire. Check with the compass to see that it is not a magnet. Magnetize it by stroking it with the magnet a dozen times. Check with the compass to see that it has become a magnet. Drop the wire several times until it becomes demagnetized.

- Remagnetize the wire, performing the compass tests as above. Hit the wire several times with the hammer. Again, it has lost its magnetic force.

- Remagnetize the wire and hold it in the flame of the Bunsen burner with the tongs. Again, it becomes demagnetized.

38. Static Electricity

Static electricity is electricity that is collected by rubbing certain surfaces together. **Electricity** is the flow of electrons. Electrons like to move about freely. As two surfaces rub together, electrons leave one and collect on the other. The surface that has all the excess electrons is **charged**. If a charged surface makes contact with the ground, then the excess electrons discharge and sparks often result. Humidity acts as a conductor and discharges static electricity most of the time. When it is very dry and cold or hot, these demonstrations work best. Students can relate to going across a carpet and getting a mild shock as they touch a doorknob. To avoid this uncomfortable experience, touch the doorknob with a key first. The same spark will happen, but the key feels no pain.

Materials: two balloons, a piece of fur or silk or nylon, string, a small fluorescent tube

- Inflate a balloon and tie it shut. Darken the room and allow time for everyone's eyes to become adjusted. Rub the balloon with the fur and touch the fluorescent bulb. Sparks will seem to move from the balloon to the bulb, then the bulb will begin to glow faintly.
- Cut a small piece of paper into very small pieces, as small as any letter in this text. Rub the same balloon with the fur and touch the paper. The paper will stick to the balloon.
- Inflate the second balloon. Hang both balloons together. Charge them both. The balloons will hang away from each other, because like charges repel and both balloons are charged with electrons. Place a student between them. The balloons will come together, as though they were kissing the student. Their charges discharge through the student.

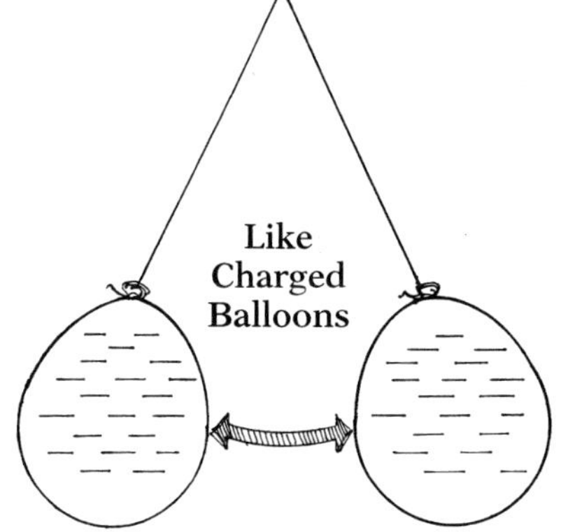

39. Static Electricity: Van de Graaff Generator

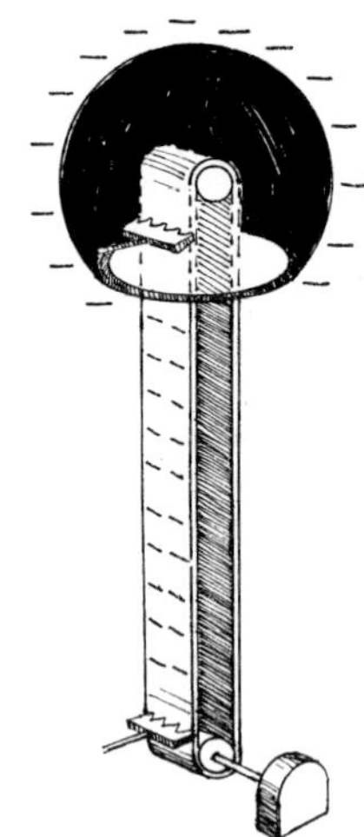

The Van de Graaff generator is a machine that generates static electricity, making it very useful in demonstrations. It consists of a small motor that turns a belt at about 3600 rpm. The rapidly moving belt touches a brush, collecting static charges, and transfers the charges to a spherical dome. In theory, the machine can generate about 5.3×10^{-9} amperes; but in practice, it makes only about 2 to 2.5 microamperes—far below the safety limit. The generator provides about 200,000 volts when its belt is 1 inch wide. The rubber belts will operate for about 20 hours, and they work well in humidity above 75%.

Materials: Van de Graaff generator, fluorescent tube (small), Ping-pong ball, stand with ring, string, adhesive tape, several small strips of silk or other conductor

- Attach the Ping-Pong ball to a string with adhesive tape. Hang it from a stand clamp. Install the ring at the same height as the dome of the generator. Position the hanging ball between the ring and the generator. Wrap one end of a piece of copper wire with stripped ends to the stand, and either tie the other end of it around a water pipe or stick it into the ground hole in an electric outlet. (**Caution: Do not connect the wire unless you are 100% sure that you have the right hole.**) Gradually move the generator closer to the ball until the ball moves; it will be attracted then repulsed and pushed to the ring. The ball will scurry back and forth.
- Darken the room and bring the fluorescent bulb close to the working generator. It will begin to glow.
- Hang a few pieces of silk above the generator and slowly bring them closer, until they all charge and start being repelled in all directions.
- To raise a person's hair, have a volunteer stand on a large block of polystyrene insulation and touch the collector with a pointed metal object. (Direct contact with fingers may be uncomfortable.) After a few seconds, you will observe the hair standing up a few strands at a time. For the most striking results, choose a person with freshly washed, light-textured hair, without any mousse or other chemicals.

40. Electricity Makes a Magnet

A wire conducting electric current has a magnetic field around it. Hans Christian Oersted, a Danish scientist, discovered this more than 200 years ago.

Materials: battery, transparent compass, coil of bell wire, stand, ring holder, two clamp holders, bell wire, a small piece of cardboard about 4 inches square, four or more small compasses, iron filings

- Place a coil of bell wire on the overhead projector. Place the transparent compass in the middle of the wire. Connect one end of the battery to one end of the coil of wire. Touch the other end of the wire to the battery so that current flows through the wire. As you touch the battery, the compass deflects sharply, indicating the presence of a magnetic field. Do not leave the battery connected for too long, because you have a short circuit.

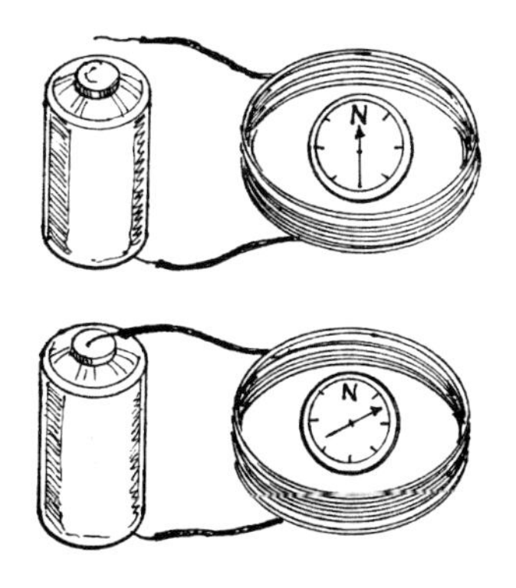

- Repeat the activity, placing a single wire across the compass and touching the battery. You will obtain the same effect as with the coil, but with less deflection of the compass. (This is Oersted's original experiment. He noticed the deflection, while others failed to observe it.)

- Punch a small hole in the center of the cardboard and pass the wire through it. The wire must be perpendicular to the cardboard. Place the cardboard on the ring. Clamp the wire at the top and bottom. Place the compasses around the wire, about 2 inches away. Connect the battery to the wire and observe how the compasses form almost a ring around the wire, indicating magnetic forces. Remove the compasses and sprinkle iron filings around the wire. With the wire connected to the battery, tap gently on the cardboard. You should obtain concentric rings.

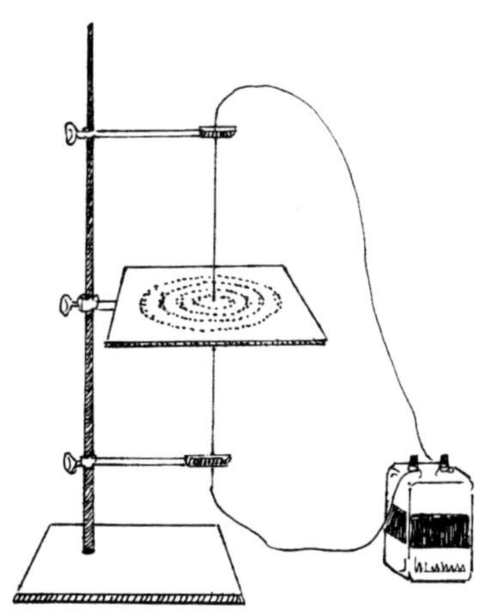

41. Electromagnets

An electromagnet is a magnet only as long as electric current goes through the circuit. The magnetism from a short piece of wire is weak, so by winding many turns and changing the wire into a coil, one gets more magnetism. Electromagnets have poles like regular magnets. Electromagnets can be turned on and off, unlike regular magnets. Electromagnets are used to pick up junk metal, in making relays, and in solenoids;they control the water valves in dishwashers and the horns in cars, they open and close garage doors, and they have thousands of other applications.

Materials: two or three batteries, bell wire, iron filings, large piece of cardboard

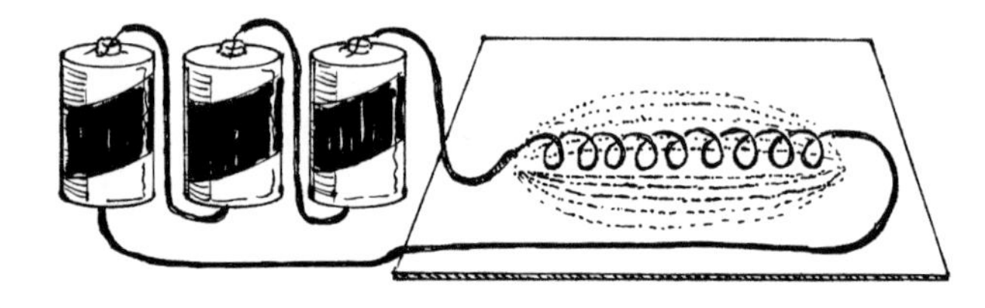

- Remove the insulation from the wire. Wrap the wire tightly around a pencil to form a long, tight coil. Sprinkle iron filings on the cardboard. Place the wire coil on the cardboard, in the middle of the iron filings. Connect one end of the wire to the negative end of the batteries in series, and connect the other end to the positive. Keep the electricity going until you get a pattern.

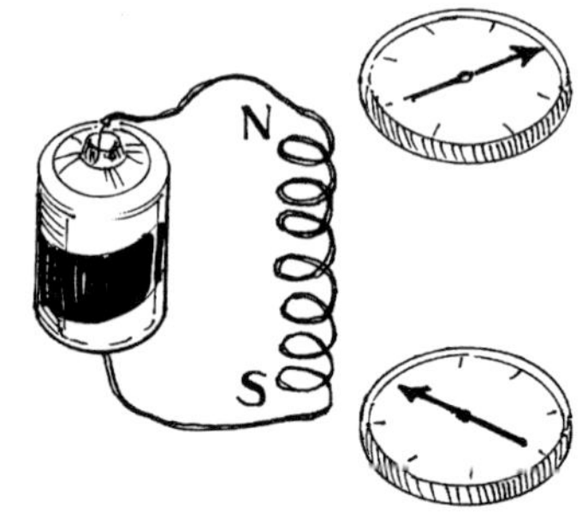

- Place a compass near each end of the coil and briefly connect the electricity to the coil. This will establish the coil's polarity. At the south pole, the compass needle points toward the coil. At the north pole, the compass needle points away.

42. Making Electromagnets Stronger

With a coil of about 20 turns, you may be able to lift one paper clip. To increase the magnetic strength of the coil, you can:

1. Increase the number of turns.
2. Place a piece of iron in the center of the coil, as its core.
3. Increase the current in the wire.
4. Any combination of these steps.

Materials: bell wire, long nail or iron rivet, batteries, paper clips

- Wrap about 20 turns of wire around the nail. Pull the nail out. Try to lift paper clips with the coil. If you can lift one, you are doing well.

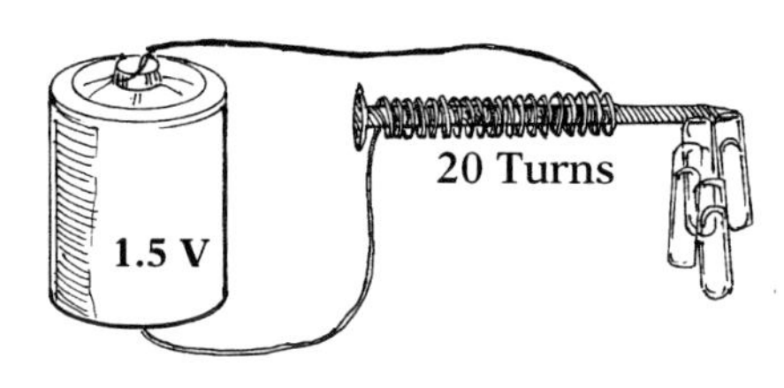

- Slip the nail back into the 20 turns of wire and turn on the electromagnet. Lift paper clips with the end of the nail. You will lift several.
- Wrap 50 turns of wire around the nail (you can start a second layer). Turn on the electromagnet and lift paper clips. You will be able to lift many more paper clips than before.

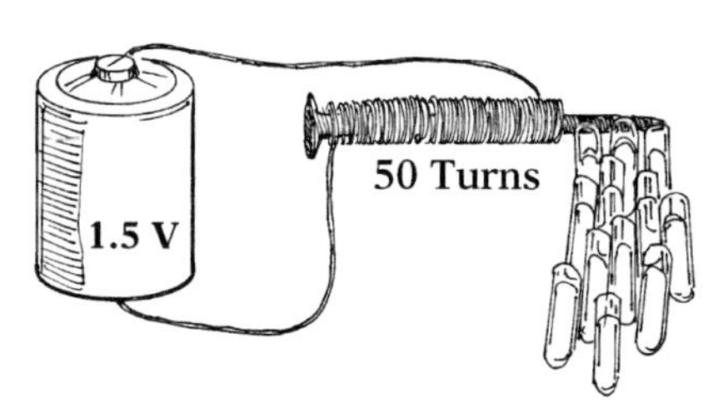

- Connect a second battery in series with the first. Turn on the electromagnet and lift paper clips. You will lift even more paper clips.

43. Inducing Electricity by Magnetism

Magnetic force is generated around a wire when electric current goes through it. Conversely, if a magnetic field cuts across a wire, an electric current is **induced** in the wire. This principle is used to generate electricity and to make all types of electrical-generating equipment. The amount of electric current increases with:

1. The number of wire turns.
2. An increase in the magnetic field strength.
3. The speed of movement of the magnetic field.

Materials: a projection galvanometer (if available) or a regular one, a coil of bell wire, a bar magnet, a larger horseshoe or more powerful magnet, a hand-crank generator

- Place the galvanometer on the overhead projector. Connect it to both ends of the bell wire loop. Holding the loop vertically, move the bar magnet forward and backward through the loop. Observe the increased reading as you move through the loop. Notice also that as you withdraw the magnet, the current appears to move in the opposite direction on the galvanometer. You have just made alternating current, or AC. Demonstrate that you can also hold the magnet still and move the coil of wire over the magnet with identical results.

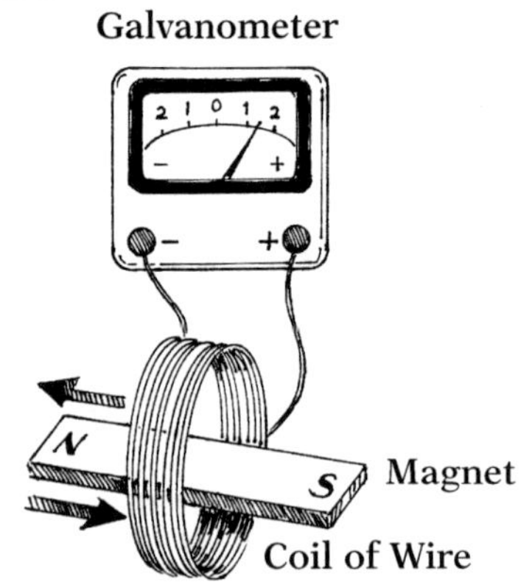

- Repeat this demonstration, moving the magnet in and out more rapidly. Notice the increase in current.
- Repeat the previous activity using the more powerful magnet. This time lay the loop flat on the projector and move the magnet up and down. Notice the increase in the electric current.
- If you have a hand-crank generator, have all students form a large ring, holding their hands together. At this point give them a mild shock. Explain to them that the generator is nothing more than a moving coil inside the magnetic field of four to six horseshoe magnets. First show how they receive no shock if they are in parallel with the bulb. Next open the circuit, put the students in series, and watch out!

Caution: If anyone wears a pacemaker or does not want to feel a mild shock, do not have them participate.

44. INDUCING ELECTRICITY: ELECTROMAGNETS

Working electromagnets can **induce** electric current in another circuit. If two coils share the same core and the **primary coil** conducts electricity, it induces magnetism in its core. When the shared core is magnetized, it produces electricity in the **secondary coil**. This is true for alternating current. With direct current, this happens only when you open and close the circuit. Electromagnetic induction is used in cars to provide the high voltage to fire the spark plugs. Electromagnetic induction is also used to step electric voltage up or down with transformers.

MATERIALS: switch, battery, stripped wire, nail, galvanometer, battery holder, wire stripper

- Wrap 30 turns of wire around the upper half of the nail and leave about 6 inches of wire at each end. Do the same on the bottom half of the nail. Connect the upper wire to the galvanometer. Connect the lower wire to the positive pole of the battery and to the switch, then wire the switch to the negative pole. Close the switch and observe the reading on the galvanometer.

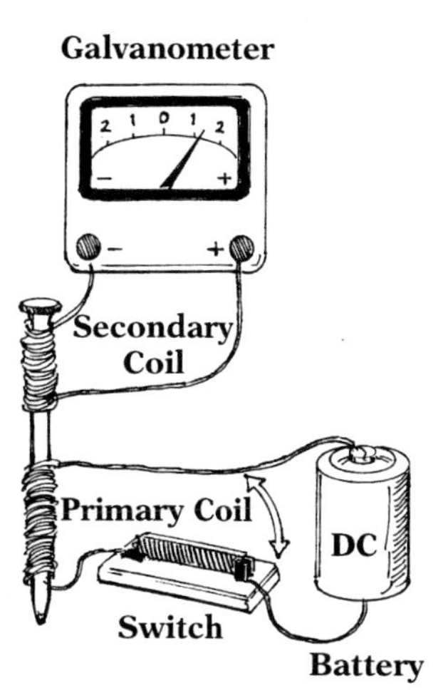

45. Series Circuits

A **circuit** is a path where electric current can flow. Circuits are open and closed by switches. Switches can be made in many ways; there are knife-blade, push-button, sliding, toggle, mercury-type, and many other kinds of switches. The simplest switch is a break in a wire; you operate it by touching the wires together or pulling them apart. Electricity can flow in a circuit only if electrons leaving the negative end of the power supply can return to the positive end. The power supply can be either direct current (DC) or alternating current (AC). The purpose of a circuit is to operate something run by electric current, such as an appliance, a motor, a light, or a TV. **Series circuit** means that electric current has only one path and must go through all the components of the circuit.

Materials: a battery, a bulb with base, a switch, and wire to connect these components

- Connect the battery to the bulb, then to the switch and back to the battery. Show how electricity has only one path to follow. This is a series circuit. The bulb lights up only when the switch is closed. If you had several bulbs in series and one burned out, the circuit would be open and none of the bulbs would work. The bulb acted as a switch. In series circuits, if any component fails, the circuit opens (does not work).

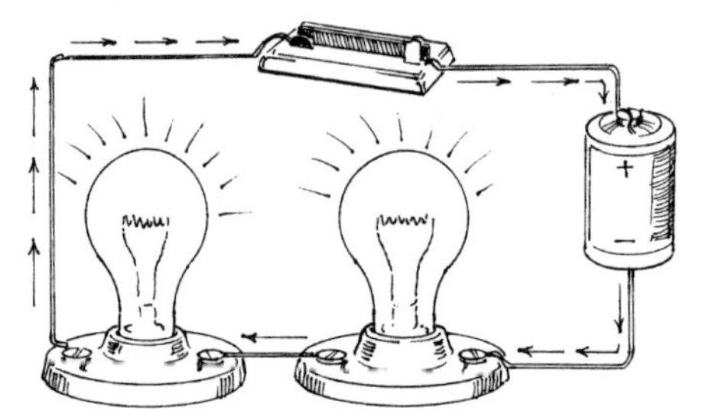

Closed Series Circuit

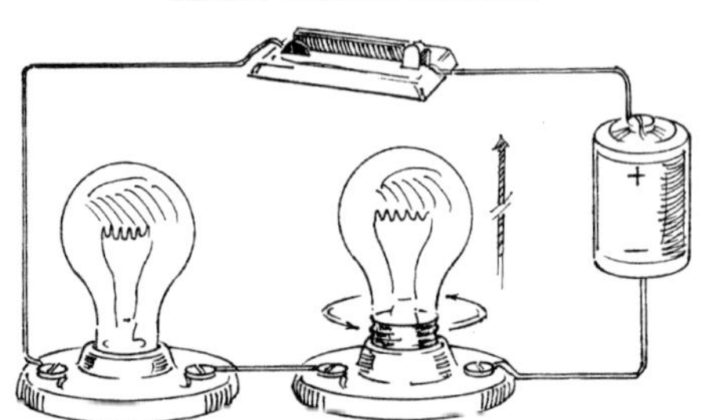

Open Series Circuit

- Connect two bulbs in series, and close the switch. Both bulbs will light up. Gently unscrew one bulb. As the bulb opens the circuit, the other bulb will also go out.

46. Parallel Circuits

A **parallel circuit** is an electrical circuit in which electrons leaving a power supply have two or more paths they can follow to go back to the power supply. If several bulbs in a lamp were wired in parallel, and one failed, the others would continue to work. Your classroom lights are wired in parallel.

Knife-Blade Switch

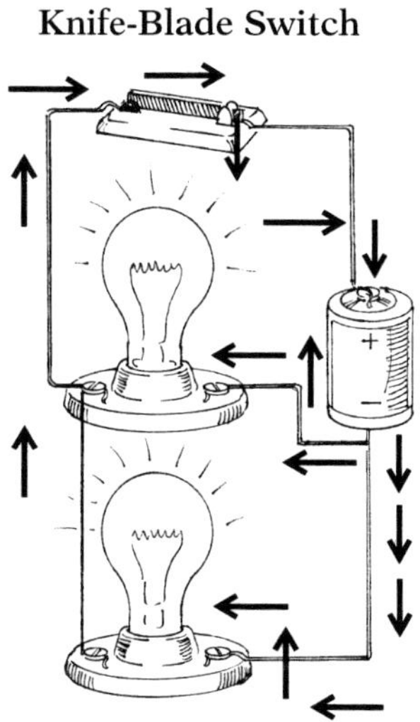

Parallel Circuit Closed

Materials: two bulbs with bases, battery with holder if needed, switch, enough wire to connect them

- Connect one end of the switch to the battery's positive pole, and connect the other end of the switch to **both** bulb bases. Connect the battery's negative pole to both bulb bases. Close the switch, and both lightbulbs will light up. Gently unscrew one bulb. The other one will remain lit and will shine a bit more brightly, because the voltage is not divided between two bulbs.

47. Short Circuits

A **short circuit** occurs when electric current from a power supply can return to the power supply without going through any resistance. A common example: Two wires of a plugged appliance touch. Short circuits are dangerous, for they start fires. To protect people, all electrical circuits have fuses, which will open the circuit if a short occurs. A new type of fuse, a ground-fault circuit interrupter, has been developed. This fuse detects minute leaks of current to ground, and by opening the circuit before a large current electrocutes a person, it saves lives. The conventional fuse trips only with large currents, by which time a person has been electrocuted. Most appliances are grounded to provide a safe return to ground of leaking electric currents. Modern electric plugs have polarized ends, with one prong wider, to provide electrical safety in case of appliance failure.

Materials: a battery, some bell wire with insulation

- Connect both bare ends of an insulated wire to a battery. Tap one end on and off and do not let it stay connected. Otherwise, in a short time you will notice that the wire is getting hot and the battery is also getting hot. Since there is no resistance in the wire, an increasingly large current tries to go through, heating the wire. The battery will heat up in trying to supply infinite current. In the process, the battery will eventually go dead.

Short Circuit

48. MEASURING VOLTAGE

A **volt** is a measure of electrical pressure and is named after the Italian scientist Alessandro Volta, who did much basic work with electricity. To measure voltage, you need a voltmeter. Electrical pressure is another term for electromotive force. A voltmeter must always be connected in **parallel** with any circuit it measures. A voltmeter can be digital (**DVM**) or analog (with a moving needle). NOTE: You can also measure the voltage of a battery, but the result is inaccurate. To get a proper reading, place the battery in its circuit and while the battery works, measure the voltage. A value of 10% below the battery's rating is fine. An alternative is to have a meter specifically designed to measure batteries.

Knife-Blade Switch

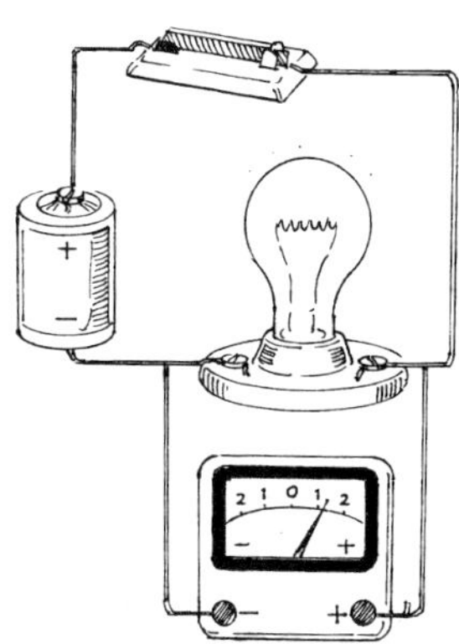

Voltmeter in Parallel

MATERIALS: a battery, a voltmeter, a bulb, a switch, optionally a small electric motor

- Read the instructions provided with the meter. Prepare a circuit, then close it and measure the voltage across the battery.

- Measure the voltage across the battery out of circuit.

49. MEASURING CURRENT

Electric **current**, the quantity of electricity going by a point in a circuit, is measured with an ammeter. The unit is the **ampere**, named after the French physicist A. M. Ampere, who did much basic work with electricity. To measure the current in a circuit, the ammeter must be in **series**.

Knife Blade Switch

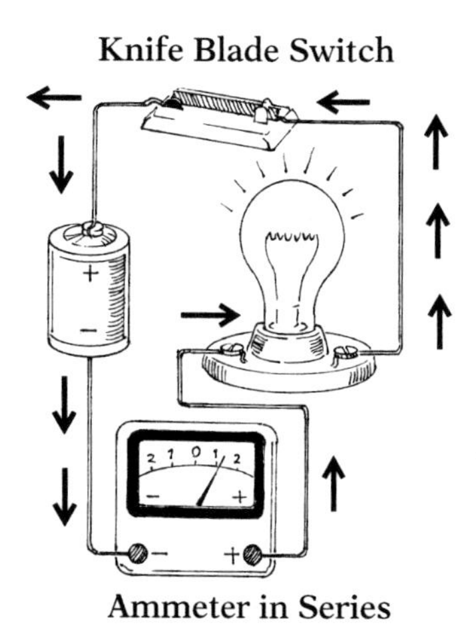

Ammeter in Series

MATERIALS: a bulb with its base, wires for connections, a battery, several resistors, and a switch

- Assemble the circuit and place the meter in series. Measure the current used by the bulb and the resistors.

50. Measuring Resistance

The **resistance** to the passage of electric current is measured with the ohmmeter and the unit is the **ohm**, named after the German scientist Georg Ohm, who did much basic work in electricity. **Insulators** oppose the flow of electric current and have a high resistance. **Conductors** have a low resistance. Factors that affect the resistance of wires are:

1. The thickness of the wire.
2. The length of the wire.
3. The material of the wire.
4. The temperature of the wire.

Wires are made from several metals. Aluminum and copper are good conductors because they have a low resistance. Toasters and electric heaters are made with wires of **nichrome**, a metal that has a higher resistance than copper and aluminum and gets hot when electric current goes through it. **Tungsten** also has a high electrical resistance and is used as the filament for lightbulbs. It gets so hot from electric current that it glows. Metals improve their conductivity if they are cooled. Materials that lose their resistance at low temperatures are **superconductors**. Mercury, a good conductor at regular temperatures, becomes a superconductor at 454° below zero Fahrenheit. Ohmmeters have internal batteries that provide power for the resistance test. Never test any circuit with power in it. Never test a battery or a live outlet for resistance.

Materials: ohmmeter and several resistors

- Measure the resistance of several resistors and compare your measurement to the posted value of the resistors.

51. Ohm's Law

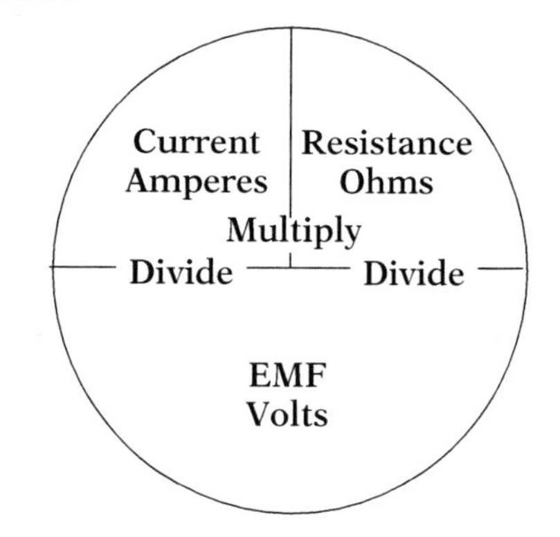

Georg Ohm figured out a basic law for electricity: $V = I \times R$.

V = Volts (electric potential)

I = Amperes (current)

R = Ohms (resistance)

If it is solved for I, then $I = \frac{V}{R}$. This provides us with much information. Current is proportional to voltage; therefore, increasing the voltage increases the current. The current is inversely proportional to resistance; therefore, as resistance increases, current decreases. Here is a convenient way to remember Ohm's law: Cover what you need and multiply or divide the other two values.

Materials: two or three resistors, voltmeter, four 1.5 V batteries, sufficient wire for connections, battery holders

- Wire the batteries through the resistor and switch. Measure the current by placing the meter in series. Measure the voltage by measuring the voltage in parallel. Take the resistors out of circuit and measure their resistances. Try the same activity, using less than 6 V. Try it with two or three batteries. Check that the measurements match the computations. Use Ohm's law to verify values.

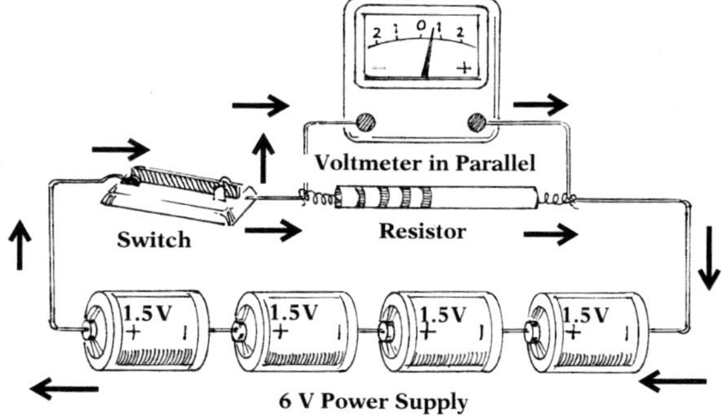

Here is a table of typical values measured and calculated.

Volts	Ohms	Amperes
6	12	.5
6	3	2
6	60	.1
4.5	100	.045
4.5	50	.09
3	300	.003
3	3	1

52. THERMOMETERS AND TEMPERATURE

Temperature is a measure of the speed of molecules. Several different units are employed to measure it. A **thermometer** is a device that measures the temperature of matter. Celsius and Fahrenheit systems are used quite commonly in the United States. Celsius and Kelvin scales are recognized worldwide. A table of temperature conversions is found in the Appendix. To find a temperature Celsius or Fahrenheit, the following formulas are used (F° is degrees Fahrenheit, C° is degrees Celsius):

$$C^\circ = \frac{F^\circ - 32}{1.8} \quad \text{and} \quad F^\circ = (C^\circ \times 1.8) + 32$$

There are 100 degrees in the Celsius scale and 180 in the Fahrenheit scale between the points where water freezes and boils. From there we get the 1.8 factor in the formula. The ± 32 is the adjustment for the freezing point in the Fahrenheit system, because 32°F is equivalent to 0°C.

MATERIALS: a thermometer, tap water, hot water, cup

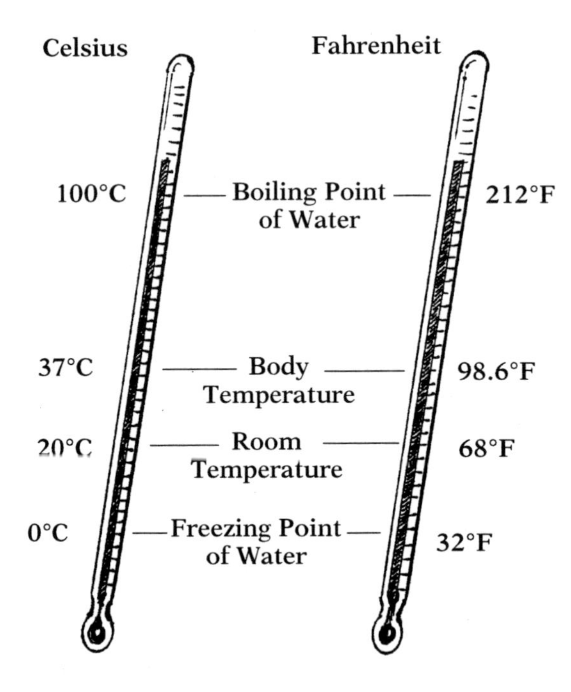

- Place some tap water in the cup and measure its temperature. Empty the cup and fill with hot water. Measure its temperature. In this demonstration, a camcorder with close-up image on a monitor is very helpful. Comment to your students that the expansion of the liquid in the thermometer is a property of heat. Solids, liquids, and gases expand when heated.

53. Calories/Calorimeter

A **calorie** is a measure of how much heat must be added or taken away to make something heat up or cool down. **One calorie raises the temperature of 1 g of water 1°C.** It would take 15 calories to raise the temperature of 1 g of water 15°C, or 15 calories would raise 15 g of water by 1°C. One kilocalorie (1000 calories) equals 1 **food calorie**. This value provides a clear way of establishing the energy in foods. See Appendix for sample values.

Materials: three pencils or dowels, masking tape, empty soda can, Celsius thermometer, water, aluminum plate, clothespin, T-pin, peanuts, matches, graduated cylinder

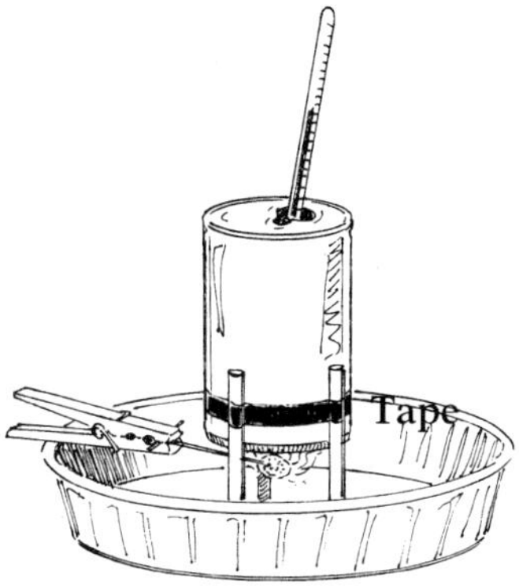

- You will measure the caloric content of a food by burning the food and using its energy to heat water. Raise the soda can on stilts, using the three pencils or dowels fastened with tape. Place 100 ml of water in the can. Stand the can in the plate. Place the thermometer in the can and measure the temperature of the water. Skewer a peanut with the T-pin and hold it with the clothespin. Light the peanut and heat the water until the flame goes out. Measure the temperature again. Multiply the temperature difference by the volume of water, and you will have the number of calories in the peanut. If you had 100 ml of water at 20°C and it heated to 25°C, then you have 100 × 5 = 500 calories or .5 food calories.

54. Force: Measuring Force

Force is a push or a pull. There are different kinds of forces; some of them are listed here:

1. The force of gravity.
2. Electromagnetic force.
3. Nuclear force.

Weight is the force of gravity; it pulls down all masses. Electromagnetic force is an electrical charge that pushes or pulls and, by moving, produce magnetic fields. It is what holds atoms and molecules together. Nuclear force is millions of times stronger than gravity, but you are not aware of it because it acts only in the nucleus of an atom. It is the strongest force known.

Materials: balance, spring scale, bathroom scale

- Measure a student's weight.
- Measure the weights of several small objects around you.
- Discuss weight as the force of gravity. How does weight differ from mass?

55. Work: Measuring Work

Work is defined as a force moving through a distance. One must use energy to do work. **Energy** is the ability to do work. You can use energy without doing work. If you push a car and it does not move, you used energy but did no work. You can have motion without work. The earth rotates around the sun, yet no work is done since there is no force pushing or pulling. The units of work have two components: the force and the distance. If you move 10 grams through a distance of 10 cm, you have work of 100 gm-cm. You can have kg-m, foot-pounds, newtons, etc. Newtons include a time component, measuring work in a unit of time. A **newton** (N) represents 1 kg moving at 1 m/sec in one second, **1 kg × 1 m/s/s = 1 N**.

Materials: a spring balance, a few objects, some string, a meterstick, a 1 kg mass

- Tie the object with the string and pull it for a chosen distance. Multiply the force as shown on the spring balance by the distance pulled, and you have calculated the work done.

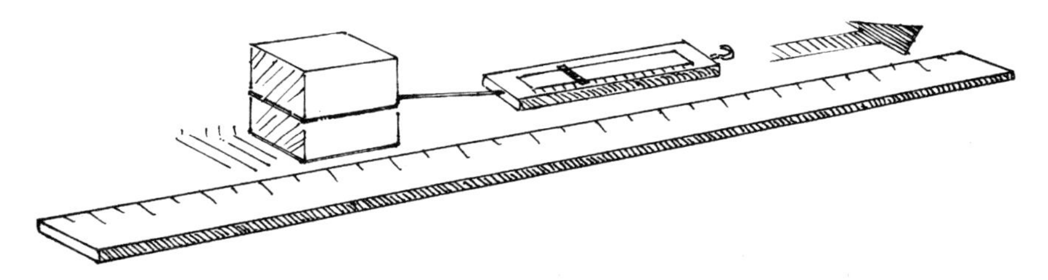

- Move a 1 kg mass over the distance of 1 m and practice until you can do it in 1 second. The force is exactly 1 newton.

56. Measuring Friction

Friction is a force that acts when one surface moves over another. Friction force opposes motion and changes motion (kinetic) energy into heat. Rubber heels are placed on shoes to increase friction, to prevent our slipping while walking. Friction has beneficial uses: Brakes on cars, bicycles, and other moving machines use friction to change kinetic energy into heat. On a bicycle, brake pads rub against the wheel's rim to stop the bike. You can reduce friction by rolling instead of dragging and by oiling the surfaces that are moving. Automobiles use oil to reduce the internal friction between moving parts. Bearings are used in moving machines to reduce rolling friction. Reducing friction saves work and energy. To **streamline** means to change the shape to offer less resistance to movement in air or water. Automobiles, planes, rockets, and ships are streamlined to offer less **drag** (friction with fluids) so that they use less fuel.

Materials: a sheet of sandpaper, a tabletop, a small piece of carpet, a mass of 500 or 1000 g, a spring balance, a piece of string

- Have students rub their hands together while squeezing them slightly together. The heat is kinetic energy lost to heat, somewhat welcome in cold climates.

- Fasten the mass to the spring balance with the string. Drag it across the sandpaper, then the tabletop surface, and again over the carpet. Notice the values on the spring balance. The force will be greatest over the sandpaper, less over the carpet, and least over the tabletop. A surface with very little friction is ice.

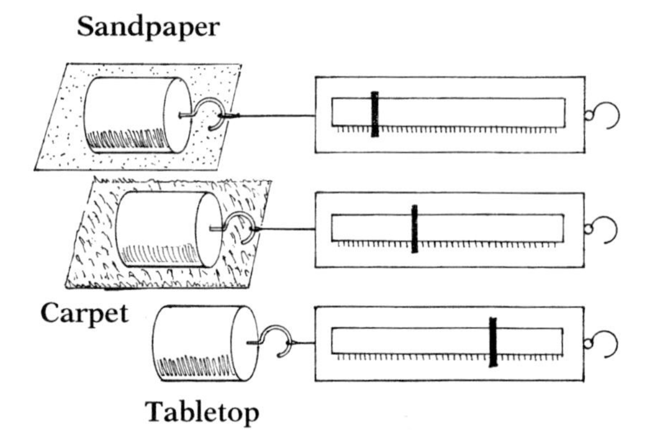

57. Center of Gravity

The place where an object is balanced is its **center of gravity**, or **c.g.** The object acts as if all of its matter is concentrated in this one spot. Balance a ruler on your finger or on the edge of another ruler. If you stand straight up against a wall and begin to lean forward without bending your hips, you soon will fall. Women can bend over much farther than men, for they have a lower center of gravity. An object tips over if its center of gravity is not over its base. Objects rotate and turn most easily about their centers of gravity.

Materials: small wooden or plastic crate, cup hook, string, rock or several metal washers, cardboard, pin

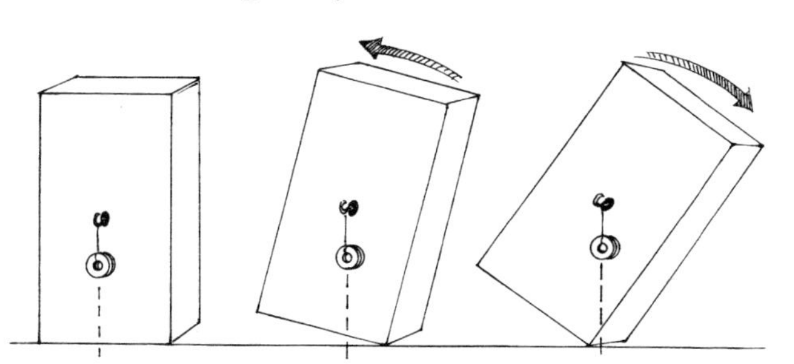

- Place the cup hook in the geometric center of the long side of the crate. Hang several washers or a small rock from it. Tilt the crate so that the washers do not hang beyond the line of the crate's base. The crate will come back on its side. Tilt the crate so that the washers hang beyond the line of its base. The crate will fall over.

- Place the box on its broad side. Shorten the string. Show that now it is fairly difficult to tip the crate over. The crate must be tipped quite a lot farther than before. Measure the distances between the c.g. and the base for both demonstrations. The closer the c.g. is to the base, the greater the stability of the object. Cars are hard to flip over because they have low c.g.'s and because manufacturers widen the distance between the tires to make their bases broader. This increases their stability against rollover.

- Cut an irregular piece of cardboard. Mark four points: *A*, *B*, *C*, and *D*. Hang the cardboard on a wall through a pin at these points, one at a time. While it is hanging at each point, attach a weight to the pin with a loop of string. Trace the line of the string on the cardboard. The four lines will converge at one point, the center of gravity. Now hang the cardboard by its center of gravity and spin it. The cardboard will spin freely. Try spinning it from any of the points *A*, *B*, *C*, or *D*. The cardboard will wobble.

58. Lever: Mechanical Advantage

Machines are devices that make work easier by placing forces where we want them, by moving things faster, and by making forces larger. A **lever** is a simple machine. The point where one pushes down is the **effort**, the pivot is the **fulcrum**, and the mass to be lifted is the **resistance**. The distance from the effort to the fulcrum is the **effort arm**. The distance from the resistance to the fulcrum is the **resistance arm**. Mechanical advantage is a number that tells us how many times a machine multiplies the effort used. Note that the work, done with or without a machine, is the same. (Actually, a machine absorbs some energy, so that it takes more energy to do work with a machine. However, the advantages of machines are so many that they are used extensively.) For a lever, the mechanical advantage is:

$$\text{M.A.} = \frac{\text{Length of effort arm}}{\text{Length of resistance arm}}$$

$$\text{M.A.} = \frac{200\text{ cm}}{20\text{ cm}} = 10$$

To find the effort, $\text{Effort} = \frac{\text{Resistance}}{\text{M.A.}} = \frac{50\text{ kg}}{10} = 5\text{ kg}$

Materials: stack of books, two pencils, crowbar, hammer, several long nails, piece of wood 2" × 4" × 4'

- Lift a stack of books slightly,using a pencil as a lever and another pencil at 90° to the first one as a fulcrum. With a minor effort you will move the books. Try moving them without the pencil lever.

- Hammer a fairly large nail partway into the wood, near one end of the board. Stand on the opposite end of the wood and remove the nail—first with the hammer, then with the crowbar. The crowbar needs much less effort, for its effort arm is longer. Repeat the same activity using the crowbar, first holding your hand close to the fulcrum, then as far away as possible.

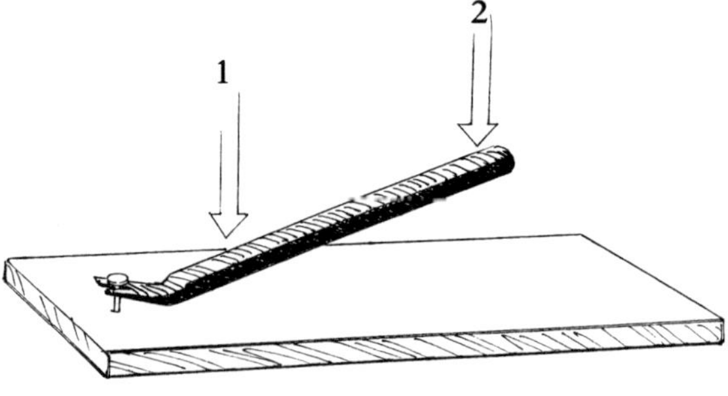

- Assemble students near a door. The hinges are the fulcrum and the door-closer is the resistance. Have one or more of them try to use one finger as far away from the hinge as possible to push the door open. Next have them repeat it, moving the hand closer and closer to the hinge. As they move the one finger closer to the hinge, they have to use more and more effort to open the door, and eventually they cannot do it at all.

59. Wheel and Axle

A **wheel** and **axle** make another simple machine. It is used to make work easier. It is a rotary lever, the lever arm being the handle of the axle. You trade turning the handle through a longer distance for the mechanical advantage. The M.A. equals the circumference of the handle's rotation divided by the circumference of the axle.

Materials: a pencil sharpener mounted at the end of a table or on a board near the end of table, string, three books, a 2–5 kg spring balance

- Remove the cover of the sharpener. Tie the string around the books and around the center axle of the sharpener. Make sure that the string is tied tightly on the shaft of the sharpener, so that it does not slip. Start cranking up the books. The effort to raise the books will be small compared to lifting the books by hand. Use a spring balance to measure the mass of the books and then pull the handle of the sharpener with the spring balance. Compare the values.

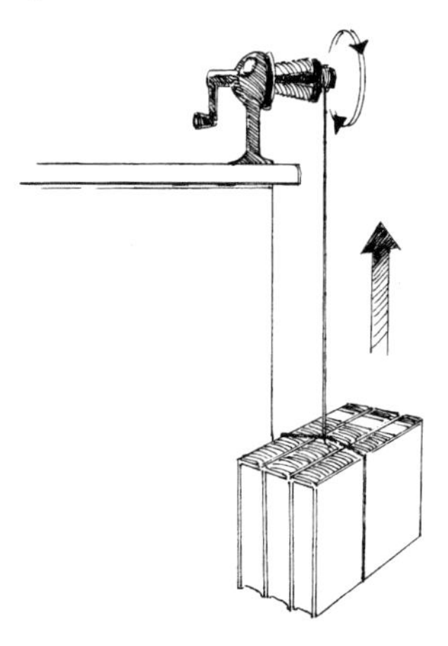

60. Teeter-Totter: Moment

A **lever** is a **first-class lever**, as are a teeter-totter (seesaw) and pliers. In a first-class lever, the pivot is between the forces. Scissors, pliers, a hammer claw, and a crowbar are all examples of first-class levers. In teeter-totters, children adjust their balance by moving closer or farther apart from the middle (fulcrum). **Moment** or **torque** is a measure of a force's ability to rotate an object about an axis (fulcrum). The weight of children multiplied times their distance from the fulcrum on a teeter-totter is a **moment**. A teeter-totter works only if the moments are balanced.

Materials: two metric rulers, string, metric masses

- Place a metric ruler between two chairs or tables. Hang another metric ruler from it. Balance it by its center. With string loops, hang two different metric masses from the ruler, one on each side, so that they balance. Their moments will be the same. (The values provided are merely guides.) If using paper clips or washers in place of metric masses, use large paper clips as hangers for both sides.

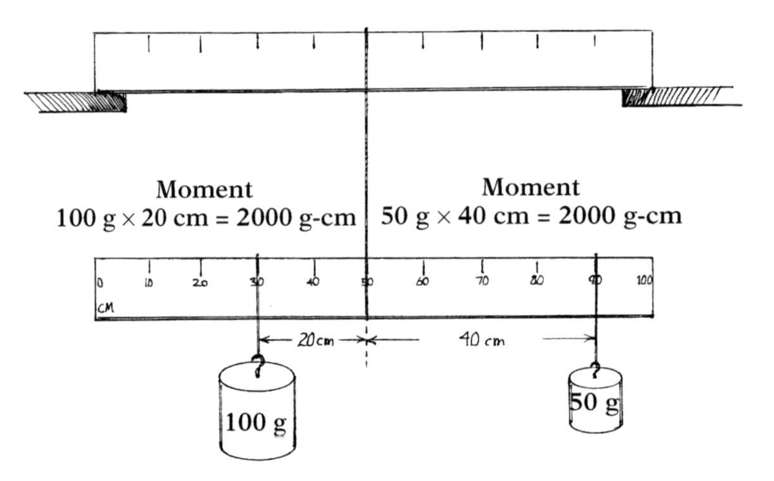

61. The Inclined Plane

An inclined plane is a ramp. Steps, roads up hills, and ramps for loading and unloading barrels on trucks are other examples of ramps. The threads of a screw are also inclined planes. The mechanical advantage of a ramp is calculated with this formula:

$$\text{M.A.} = \frac{\text{Length}}{\text{Height}} = \frac{5\text{ m}}{1\text{ m}} = 5$$

You need to note that the ramp made the rolling of the barrel easier, for you need only 10 kg of force. The final work is still 50 kg-m (10 kg × 5 m). The ramp helps us use less force over a greater distance.

Materials: an inclined plane or board of wood, a toy automobile, a spring balance

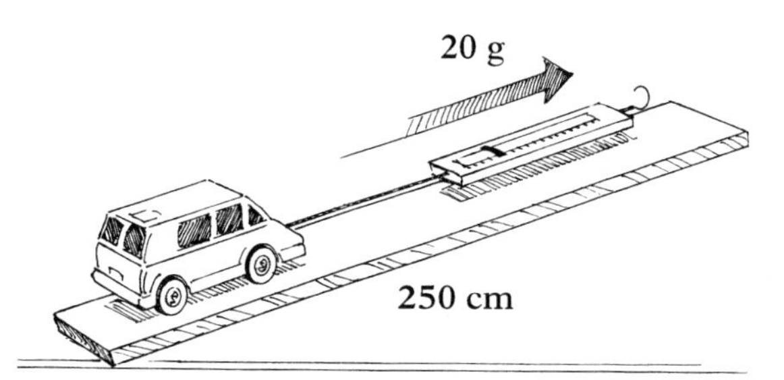

- Lift the toy auto with the spring balance to measure the force needed. Place the car on an inclined plane and measure the effort needed to pull it up. The effort will be considerably less than lifting the car straight up. You are trading the lower force for a longer distance. If the car needs 50 g to move 100 cm, the work is 5000 g-cm. If you wanted to pull it up a ramp using 20 g of force, you would need a ramp 250 cm long, for the work would still be 5000 g-cm.

62. Pulleys

A **pulley** is a wheel that turns on an axle. It is a simple machine that can change the direction of a force. Pulleys can be **fixed** or **movable**. A clothesline has two fixed pulleys on its ends. A block and tackle is a movable unit made with several pulleys. It is used to lift heavy objects like car engines.

Materials: two small pulleys (same size), two double pulleys (same size), string, five books of equal mass, three cup hooks, a piece of wood 2" × 4" × 3' or several lab stands

- Install the cup hooks on the board near its center and about 6 inches apart from each other. Place the board between two chairs or tables. Suspend one pulley from a hook. Pass some string through the pulley's groves and attach a book to each end of the string. The books will not move, for they are balanced. The fulcrum is the axle and the arm is the radius of the pulley. Since both sides are equal, there is no M.A. A single pulley is used to change only the direction of force. Lift or lower a book. Both travel at the same speed.

- Attach the assembly of two pulleys. One hook will actually support two books. The top pulley is a fixed pulley, the bottom pulley is a movable pulley.

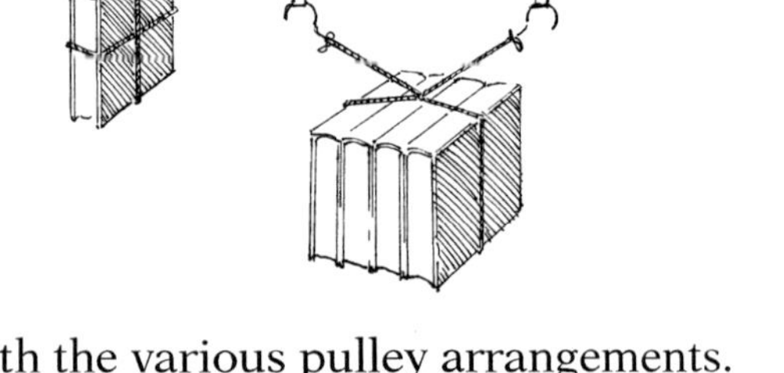

- The same activity can be extended by using four pulleys and obtaining a greater mechanical advantage. (Tackles usually have multiple gears side by side.) Attach the two multiple gears as in the illustration. Note that both upper and lower pulleys share the same axle. The upper gear system is hung by one hook, the lower one supports the books. You can also vary this activity by using spring balances to measure the force necessary to lift the books with the various pulley arrangements. This particular activity illustrates clearly the concept of mechanical advantage.

63. FRICTION AND MACHINES

Machines are made up of many parts that move. Moving parts rub against each other, and the rubbing causes friction. **Friction** force not only opposes the moving force, but also changes moving energy into wasted heat. Additionally, friction causes moving parts to wear out. Since friction wastes some of the moving energy, the mechanical advantage is always less than the **ideal M.A.** The **actual M.A.** is obtained by dividing the resistance by the effort. The **efficiency** of a machine is obtained by dividing the actual M.A. by the ideal M.A.:

$$\text{Efficiency} = \frac{\text{Actual M.A.}}{\text{Ideal M.A.}}$$

The efficiency of a machine is usually expressed as a percentage. The friction in most machines is kept to the lowest possible point through oiling. In automobiles, the engines are continuously lubricated with oil. Smaller appliances with many moving parts, like sewing machines and bicycles, require constant oiling. Machines are not ideal, so they put out less work than is put into them. The difference is friction.

MATERIALS: pulley, string, book(s), spring balance, hook, wood board

- Tie one end of the string to the hook, thread the other through the pulley and tie it to the spring balance. Using the pulley, lift up one or more books. Measure the force needed to lift the books. In the illustration, the ideal M.A. is 2. The actual M.A. $= \frac{4 \text{ kg}}{2.2 \text{ kg}} = 1.8$. The efficiency of this machine, then, is $\frac{\text{Actual M.A.}}{\text{Ideal M.A.}} = \frac{1.8}{2} = 0.9 = 90\%$. In this situation, 10% of the work is lost to friction.

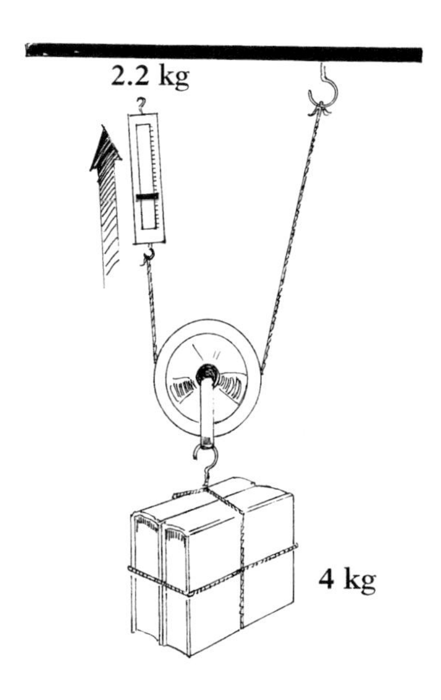

64. GRAVITY

About 300 years ago, Isaac Newton discovered the laws of gravity. Gravity explains why things fall and why the earth remains in orbit around the sun, why the moon orbits the earth and what governs the entire orderly movement of objects in the universe. Newton stated that all masses in the universe have a mutual force of attraction. This attraction increases directly with the size of their masses and decreases rapidly with the distance between them.

Newton's formula is: $F = G \frac{M \times m}{d^2}$

F is force on a mass
M is one mass
m is the other mass
d is the distance between the masses
G = a gravitational constant: $6.7 \times 10^{-11} \text{N m}^2 \text{kg}^{-2}$.

Gravitational forces are negligible unless one of the masses is very large. Since the force depends on the inverse of the distance squared, distance is also a big factor. At 3 meters, the force is 1/9; at 5 meters it is 1/25; and at 50 m it is 1/2500. Since earth has a large mass, it has a large gravitational force that attracts everything on it or near it. A space shuttle is in a free fall in space and would fall away from earth into the depths of the universe were it not for the earth's gravitational pull, which keeps it in orbit. Astronauts adjust their orbital speed so that their momentum matches earth's gravity. If they were to travel faster, they would escape earth's pull and go into deep space, as many interplanetary space probes do. If they were to slow down, they would begin to come back toward earth.

MATERIALS: a piece of chalk, a roll of masking tape, a couple of feet of string, several small objects

- Take any small object and let it fall. Explain that all objects on or near a planet fall toward it. Throw a piece of chalk or other small object up. It will eventually fall down. As it moves up, gravity slows it down until it stops. Then it begins to free-fall. 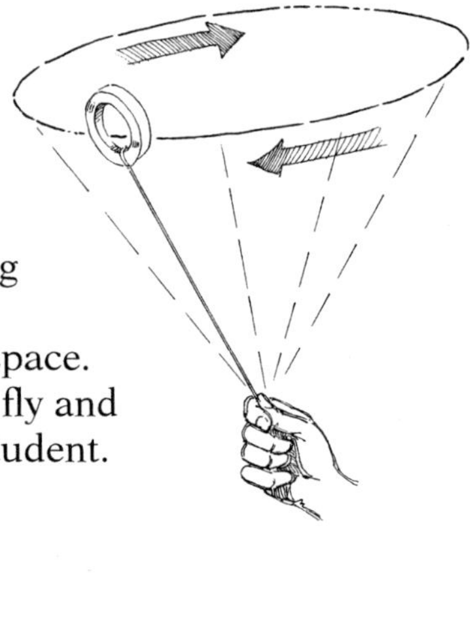
- Take a roll of masking tape or other small object and tie it to a piece of string about a foot long. Spin it around your arm in a circular orbit. Explain to your students that this string is a visible model representing the force of gravity. If gravity ceases or you speed up, the object will continue falling in space. Let go of the string and let the masking tape fly and fall somewhere where it will not harm any student.

65. Acceleration by Gravity

Earth's force of gravity attracts all objects toward its theoretical center. Surveyors and contractors use plumb bobs, knowing that they will always provide a true vertical line. Air and liquids, which create friction, or **drag**, slow down all falling objects. Falling objects continue to accelerate (change their speed) due to the force of gravity, until their drag with the air or liquid equals their own weight. Their maximum final constant velocity is **terminal velocity**. In a vacuum, heavy and light objects fall at the same speed. Galileo (1564–1642), an Italian scientist, proved that all objects fall the same distance in the same time.

Materials: several coins of different sizes, several small objects, two sheets of paper, a ruler, a deck of playing cards

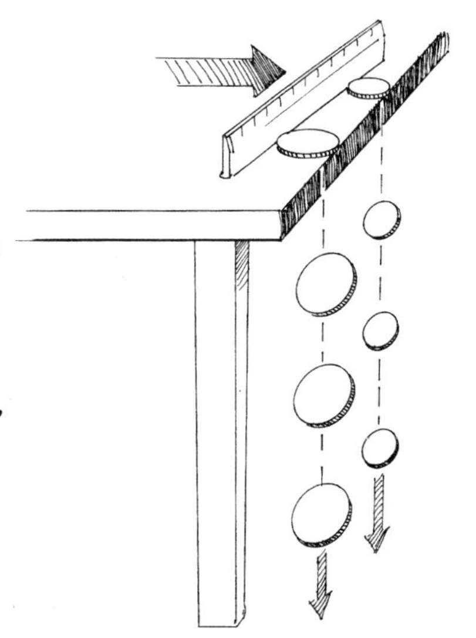

- Place two different coins near the edge of a table. Push them over the edge carefully, using the ruler, so that both start falling at the same time. Check to see if they land at the same time. They will land at the same time because the difference in their masses does not matter.

- Crumple a piece of paper into a small, tight ball. Drop it at the same timc as a flat sheet of paper. The flat piece of paper will fall more slowly, due to its greater drag with air. Modern automobiles, jets, and rockets are streamlined to reduce drag.

66. Weight

Mass is the amount of material in an object. The mass of an object does not change, except in a special case. Albert Einstein predicted that the mass of an object will increase if the object travels very fast, at or near the speed of light (300,000 km/sec or 186,000 miles/sec). (If light could bend and travel around our earth at the equator, it would make 7.5 turns in 1 second.) This special effect begins if an object moves at least 80,000 km/sec. Scientists have succeeded in speeding up electrons and other small particles. They have also been able to measure the increased mass of these speeded-up particles. **Weight** is the pull of gravity on an object. If an object has a mass of 1 kg, earth's gravity pulls on it with a force of 1 kg. Different objects, even if they have the same shape and volume, can have different weights.

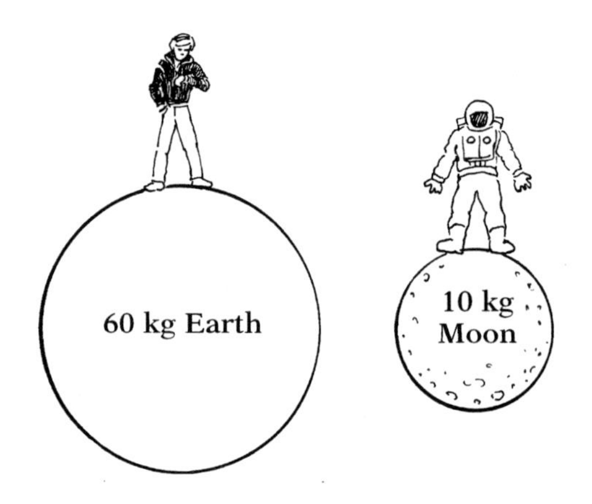

Some materials are heavier than others. If the force of gravity changes, so does the weight. On earth, objects appear to have less weight as they move farther from sea level. This is due to the increase in distance from the earth's center. The moon has one sixth as much gravity as earth. An astronaut on the moon weighs only one sixth as much as on earth. In orbit, objects are weightless. This can cause serious health problems for humans traveling in space. Their muscles become flabby and they rapidly lose many vital minerals, such as calcium.

Materials: a metric balance, a metric mass of 500 or 1000 g

- Place the metric mass on the balance and balance it. The balance measures weight. A 1 kg mass should have the weight of 1 kg. If it does not, calibrate the balance. (To calibrate means to adjust it so that it reads zero at the start.)

67. PRESSURE

Pressure is a force over a limited amount of surface. Some of its units are kg/cm^2, $lb/inch^2$ or pascals, an S.I. Unit. At sea level, pressure is also measured as a column of mercury 760 mm high. If you press your hand on an ice block, it does not break. If you use an ice pick, it concentrates the force in a very tiny area, and the resulting pressure breaks the ice. If you stand on the floor, only part of your shoes is in contact with the ground. If 10 square inches of your shoes are in contact with the floor and you weigh 150 lb, then the pressure on the floor is 15 lb/in^2.

$$\text{Pressure} = \frac{\text{Force}}{\text{Area}}$$

A sharp knife has less area in contact with an object than a dull knife. By making the contact area very small, the pressing force acts as if it has been increased. This is why a sharp knife cuts more easily than a dull one.

MATERIALS: a pencil, a Magdeburg sphere with vacuum pump

- Have your students hold a pencil and press the tip against the palm of their other hand. They will feel a sharp point with pressure. Have them repeat this activity by turning the pencil upside down so that the eraser is against the palm. The pressure will appear to be less, for the force is concentrated over a greater area.

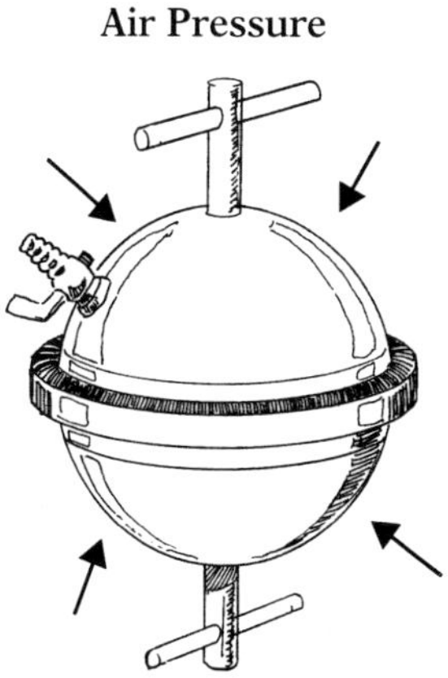

- Using a vacuum pump, evacuate the air from the Magdeburg sphere. Ask individual students to try to take the hemispheres apart, without opening the inlet air valve. A very large force will be needed to accomplish it. You may wish to compute the surface area of the sphere and the pressure on it.

Air Pressure

Magdeburg Sphere

CAUTION! Have all students trying to pry apart the sphere stay well away from others, in case the sphere comes open. Have them also stay clear of tables and walls.

68. Measuring Pressure

Pressure is measured with a **manometer**. A manometer is a U-shaped tube filled with colored water. Some manometers are filled with colored oil. Other pressure gauges are similar to aneroid barometers. Pressure gauges are used to measure tire pressure and oil pressure, among other uses in automobiles. In aviation, pressure is used to measure flight speed and for the vital fluid pressures in the aircraft and its power system. In early aviation, pressure was used to find altitudes. Pressure gauges are used on air compressors and most pressure-operated devices. Pressure gauges are also used by sea divers to measure their depths. The high pressures of oceanic depths cause nitrogen (75% of the air we breathe) to dissolve in the diver's blood system. If divers come up too rapidly, they will experience the **bends**, pain, and possibly death. The nitrogen will bubble in the circulatory system. To allow for the dissolved nitrogen to leave the body, divers must surface slowly. Then the nitrogen returns slowly to the lungs and is exhaled.

Materials: glass tubing about 18 inches, 3 to 4 feet of rubber hose to fit the tubing, Bunsen burner, hot-melt glue gun, glue sticks, water, food coloring, 6" × 6" board, 6" × 10" board, 1" × 1" × 12" support frame, thistle funnel or other small funnel to fit into the rubber hose, rubber sheeting, rubber band, fish tank

Manometer Assembly: Bend the glass tube, using a Bunsen burner. Try to obtain a shape similar to a letter U, with a 2" to 3" horizontal section. Fasten the assembly to the board with some hot-melt glue. Let the end where you will connect the rubber hose extend slightly off the board. Attach this board to the support wood with the glue and fasten the support piece to the base. Fasten one end of the rubber tube to the glass tubing and the other end to a thistle funnel or other small funnel. Using colored water, fill the glass tubing up to about 1 cm below the rubber tubing. Stretch the rubber membrane over the thistle funnel and fasten it with a rubber band, stretched over itself several times. Make reference marks on the opposite end of the manometer.

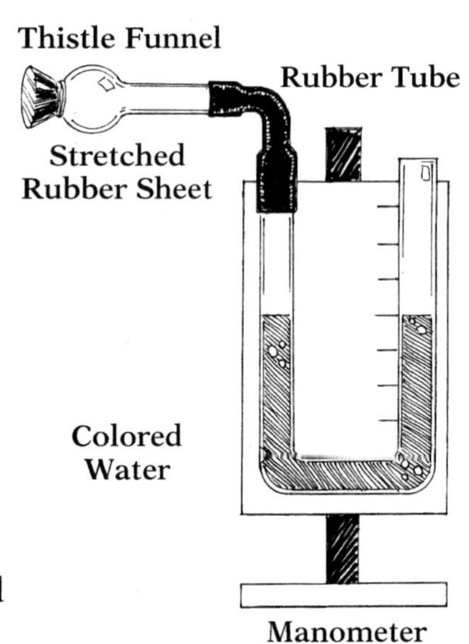

Manometer

- Press on the rubber membrane with your thumb. You will notice a rise in pressure.
- Place the thistle funnel inside a fish tank filled with water. Show the increase of pressure with depth.
- Demonstrate that at a specific depth there is no change in pressure. Let the funnel lie on the bottom of the fish tank and turn it in several directions.

69. Density and Specific Gravity

Density is a physical property of matter. It tells how much of it there is in 1 cubic centimeter. The density of water, a standard, is 1 gram per cubic centimeter. If objects of the same size have different masses, it is because they have different densities. The density of an object is found by dividing its mass by its volume. (The next activity, Archimedes' Principle, addresses how to find the volume of irregular objects.)

$$\text{Density} = \frac{\text{Mass}}{\text{Volume}}$$

Following are the densities of a few familiar materials.

Table of Densities

(Also see Appendix)

Material	Density g/cm3
Gold	19.3
Mercury	13.6
Lead	11.3
Copper	8.9
Brass	8.5
Iron	7.9
Aluminum	2.7
Sugar	1.6
Corn syrup	1.1
Water	1.0
Salad oil	0.9
Wood (Elm)	0.8
Alcohol	0.8
Cork	0.2

Specific gravity is nothing more than the density of a material without its units. This is accomplished by dividing the density of a material by that of water. The numerical values remain the same while the units divide out.

MATERIALS: two beakers, salad oil, white vinegar, corn syrup (dense, clear), blue and yellow food coloring, water

- Place a beaker on your overhead projector, for bottom illumination. Inform your students that you are about to make a salad dressing. Ask them to name the various ingredients needed. Pour a little salad oil in the beaker. Add the same amount of vinegar, to which you have added a drop of blue coloring for better visibility. Add about the same amount of corn syrup. You will obtain a beautiful separation of substances, due to their different densities. Pour the dressing back and forth a few times between the two beakers, then let it stand. The substances will separate again by their densities. Ask the class where water would go if it were added to this salad dressing. Then slowly pour in the same amount of water, after giving it a drop of yellow coloring. It will mix with the vinegar, changing the blue to green.

70. Archimedes' Principle

About 2000 years ago Archimedes, the Greek scientist who developed **Archimedes' principle**, lived in Syracuse, Sicily. He observed that as he floated in a bath, he appeared to be lighter, and that the level of water rose as he entered the tub. He discovered that as a body is placed in water, it displaces its own volume of water. This finding allows us to find the volume of any irregular body by dipping it in water. Archimedes also noticed that the weight of the liquid that was displaced was the same as the apparent loss in weight of the item placed in water. This is Archimedes' principle.

Materials: overflow dish, graduated cylinder, small rock, spring balance, string

- Hang the rock by a string from the spring balance and measure its mass. Then submerge the rock in the water and measure its mass. Measure the volume of displaced water. Then compare the volume of displaced water to the apparent loss of mass of the rock when immersed in water. They should be nearly equal, for 1 cm^3 of water has a mass of 1 g.

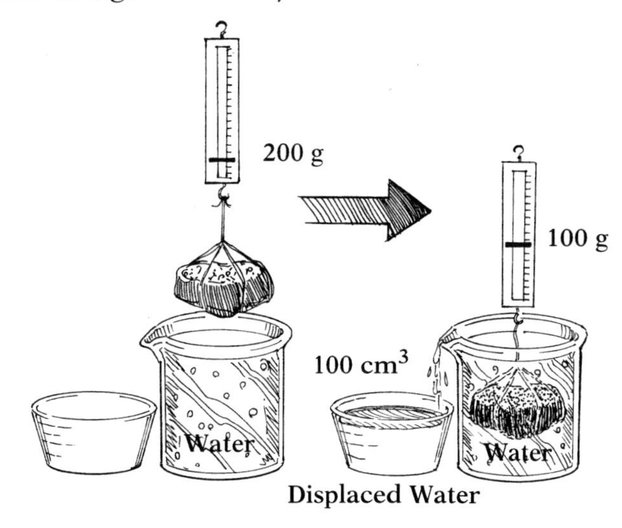

Displaced Water

Applied Conclusion: When you put an object in water and it appears to become lighter, the difference between its mass in air and in water represents its volume in cm^3. No overflow dish is really needed, except for the original proof.

71. Making Sounds

Sound is caused by objects vibrating back and forth. With a tuning fork, you can make a sound of only one pitch (frequency).

Materials: tuning fork, glass nearly full of water, rubber mallet if available, record player, old record, index card, sheet of paper, two metal pins

- Hit the tuning fork with the rubber mallet. If a mallet is not available, hit the tuning fork on the rubber heel of your shoe. Hitting the fork on a hard material causes the fork to lose its calibrated pitch (frequency). Point the fork toward the class and repeat several times.
- Touch the handle of the vibrating fork to a student desk and have the students listen by placing their ears on the surface. Repeat for the entire class.
- Since it may be difficult to see that the tuning fork is really vibrating, place the vibrating fork into the glass of water. It will create a gentle spray. The water acts as a dampener (reducer) of energy. The tuning fork will quickly cease to vibrate. The water behaves like a car's shock absorbers or a piano's dampener: In each of these, kinetic energy is transferred into another material by doing work (pushing water or oil or piano strings).
- Place the record on the record player. Make a player needle by pushing one pin through an index card. Holding the card gently, place the needle in the groove of the record. You will be able to hear a faint sound from the record as the card vibrates in tune with the needle. Thomas Alva Edison invented the phonograph, the "talking machine" of his day. If you need more sound, make a paper cone, fold over the closed end, and place a pin through it. Again, place the needle in the record groove. Louder sound will be heard. The paper acts as a mechanical amplifier.

72. Loudness and Sound (Decibels and Amplitude)

Decibels are units of measurement for the loudness of sounds. They are logarithmic; this means that they are not linear. Any change of ±3 dB means a doubling or halving of the loudness of sound. Since sound is transmitted through material objects as well as air, decibels measure the pressure of the sound waves. Since all sounds are made by vibrating objects, those that have more energy appear to be louder. When a person is exposed to loud sounds, the eardrum protects itself by growing thicker, an irreversible process known as deafness. People exposed to loud sounds must wear protective ear devices. Exposure to 90 dB can cause temporary deafness. Electronic supply stores carry inexpensive yet accurate decibel meters, which you can use to check the level of sounds to which you are exposed. Use one to check sound around you and during school events like dances.

Table of Sounds in Decibels

Sound Pressure Decibels	Sound Source
0	No sound
15	Rustling of leaves
40	Whisper
45	Normal household
65	Normal talk
75	Busy street traffic
85	Vacuum cleaner
100	Subway, or riveting
110	Thunder
115	Rock band near
120	Painful to ears
160	Jet engine
200	Rocket takeoff

Materials: toy xylophone, whistle, tuning fork, radio, other sound makers

- Play the various sound-making devices. Finally, play anything on the radio and adjust the volume from very low to fairly loud and back down again.

MATERIALS: tuning fork, cathode ray oscilloscope (CRO), microphone, radio

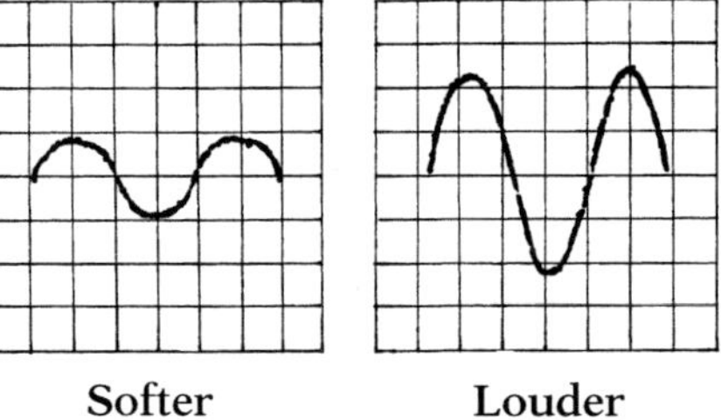

- Hook up the microphone to the CRO. Show the waves of sound caused by the tuning fork. Show the waves of a faint music program. Repeat with a greater volume. The waves will appear to be higher. The height of a wave is its **amplitude**, and it is the measure of a wave's energy—in our case of its loudness.

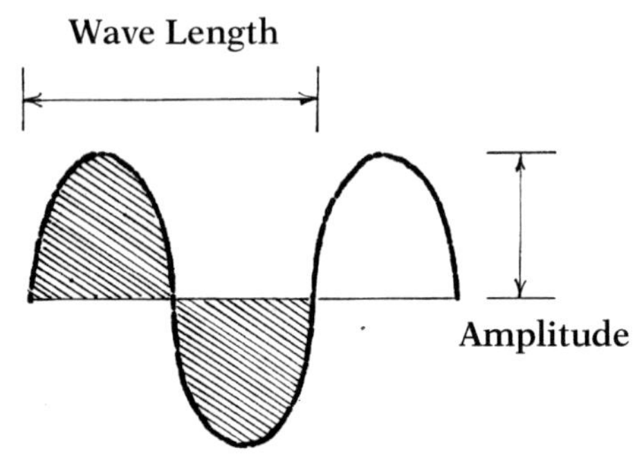

73. Pitch (Frequency) and Sound

Vibrations that cause sound are between 20 and 20,000 vibrations per second. This is the **audio spectrum**. The number of vibrations per second is **frequency**. The unit of frequency is the **hertz (Hz)**, after Heinrich Hertz, who did much pioneering work that led to the invention of radio. Sounds above the audio spectrum, too high to be heard by humans, can be heard by dogs and many other animals. This explains why certain dog whistles are soundless to humans.

Materials: three tuning forks of different frequencies, a small radio with bass and treble adjustments, optional sine-square wave generator, speaker, and cathode ray oscilloscope (CRO)

- Play each tuning fork separately, starting from the lowest and going to the highest. Point out that as you increase the pitch (frequency), the sound gets higher in frequency. Each fork will have stamped on its body its vibrating frequency.

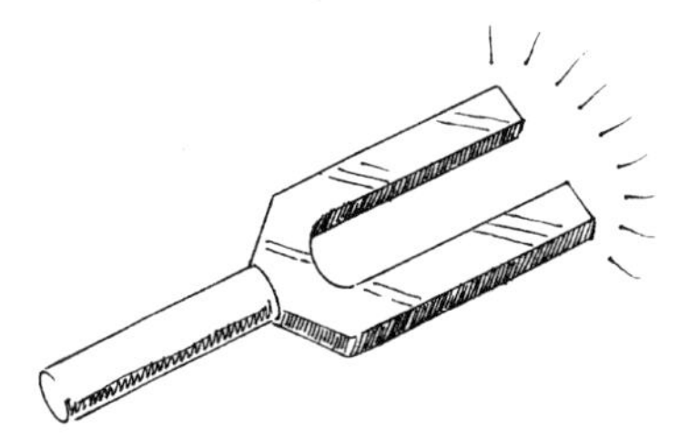

- Play some music on the radio. Turn the bass control up and down, then repeat with the treble control. Students will immediately comprehend higher and lower pitch.
- Connect the sound generator to the speaker and to the CRO. Go from low- to higher-pitched sounds. Students will not only hear sound differences clearly, but also be able to observe how the number of waves increases as you increase their frequency.

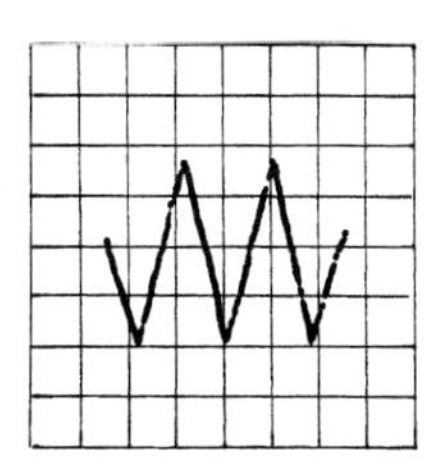

Lower Frequency

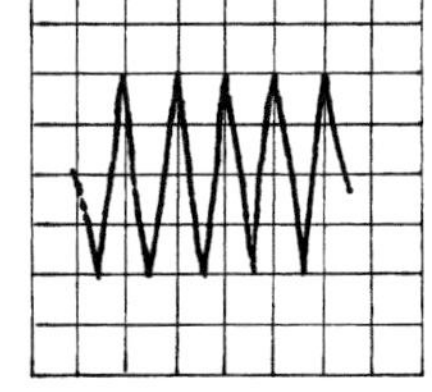

Higher Frequency

74. Transmission of Sound Through Materials

Sound is transmitted through material objects (solids, liquids, and gases) because sound waves cause compression and rarefaction (expansion) in the distance between the molecules of matter. Where there is no matter, as in a vacuum, then sound is not transmitted. Sound eventually stops, because its energy is used up in moving the molecules of matter. Sound is transmitted by either **compression** or **transverse waves**. The speed of sound is about 335 meters per second in air, 1500 meters per second in water and about 5000 meters per second in metals. Since sound travels at a fraction of the speed of light, if you count the time between seeing a bolt of lightning and hearing the thunder, you will know how far away the lightning has struck. Sound travels 1 kilometer in 3 seconds.

Materials: slinky, tuning fork, glass of water

- Have one student hold the Slinky about 20 feet away from you. Shake it up and down and show the travel of a **transverse wave**. You may wish to place it on a flat table and shake one end. If you shake while a wave is returning, the two waves will collide. Waves can **aid** (help each other) or **buck** (cancel their energies).

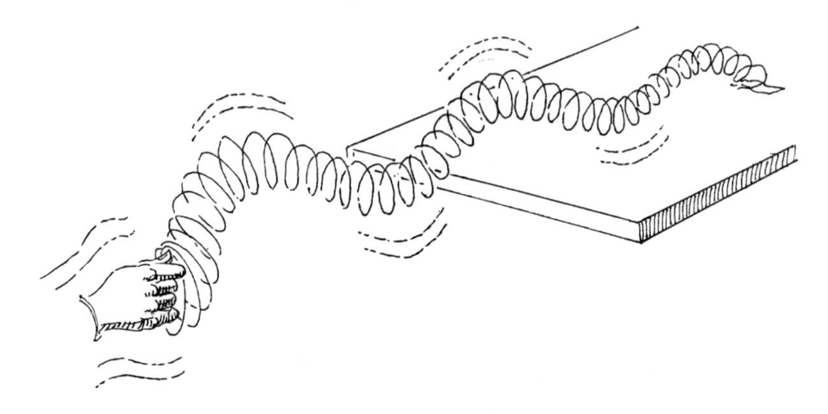

- While holding the Slinky, push rapidly along the length of the spring, not laterally or up and down. You will notice that there is a small region of compression of springs that travels forth and back. That is a **compression wave**. A small amount of transverse wave is unavoidable. These two activities will work equally well on earth and in space.

- Generate a transverse wave and hold one end of the Slinky to a student's ear. Pluck the Slinky. It will make many interesting sounds.

75. Resonance

Most sounds are made up of a combination of several frequencies. A tuning fork makes a pure sound, for it produces only one frequency. The particular frequency at which an object vibrates is its **natural frequency**. When you have two objects with the same natural frequency, one will vibrate when the other one does. This phenomenon is **resonance**. Resonance explains all radio transmissions. Your radio has an electronic tuning fork that adjusts to the frequencies of the transmitting stations. As you tune in a particular frequency, the radio station signal is heard.

Materials: two boxes (open on one end) with identical tuning forks mounted on top, rubber mallet (resonance apparatus)

- Place the openings of the boxes so that they face each other a few inches apart. Hit one tuning fork. Observe how the other one responds by vibrating. Since a small amount of energy will be transmitted, have students listen by placing the ear on the edge of the box. Change the frequency of the fork by sliding the adjustable sleeve up or down. Hit one fork and adjust the other fork until it is tuned to thc first one.

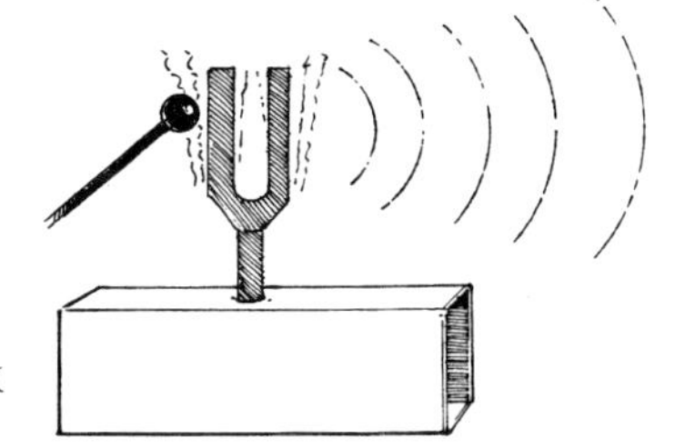

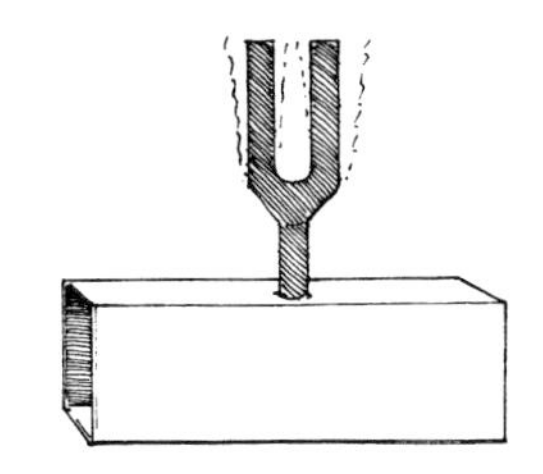

Appendix

1. Temperature Conversion (Celsius to Fahrenheit)

C°	F°	C°	F°	C°	F°	C°	F°	C°	F°	C°	F°
250	482.00	**200**	392.00	**150**	302.00	**100**	212.00	**50**	122.00	**0**	32.00
249	480.20	**199**	390.20	**149**	300.20	**99**	210.20	**49**	120.20	**-1**	30.20
248	478.40	**198**	388.40	**148**	298.40	**98**	208.40	**48**	118.40	**-2**	28.40
247	476.60	**197**	386.60	**147**	296.60	**97**	206.60	**47**	116.60	**-3**	26.60
246	474.80	**196**	384.80	**146**	294.80	**96**	204.80	**46**	114.80	**-4**	24.80
245	473.00	**195**	383.00	**145**	293.00	**95**	203.00	**45**	113.00	**-5**	23.00
244	471.20	**194**	381.20	**144**	291.20	**94**	201.20	**44**	111.20	**-6**	21.20
243	469.40	**193**	379.40	**143**	289.40	**93**	199.40	**43**	109.40	**-7**	19.40
242	467.60	**192**	377.60	**142**	287.60	**92**	197.60	**42**	107.60	**-8**	17.60
241	465.80	**191**	375.80	**141**	285.80	**91**	195.80	**41**	105.80	**-9**	15.80
240	464.00	**190**	374.00	**140**	284.00	**90**	194.00	**40**	104.00	**-10**	14.00
239	462.20	**189**	372.20	**139**	282.20	**89**	192.20	**39**	102.20	**-11**	12.20
238	460.40	**188**	370.40	**138**	280.40	**88**	190.40	**38**	100.40	**-12**	10.40
237	458.60	**187**	368.60	**137**	278.60	**87**	188.60	**37**	98.60	**-13**	8.60
236	456.80	**186**	366.80	**136**	276.80	**86**	186.80	**36**	96.80	**-14**	6.80
235	455.00	**185**	365.00	**135**	275.00	**85**	185.00	**35**	95.00	**-15**	5.00
234	453.20	**184**	363.20	**134**	273.20	**84**	183.20	**34**	93.20	**-16**	3.20
233	451.40	**183**	361.40	**133**	271.40	**83**	181.40	**33**	91.40	**-17**	1.40
232	449.60	**182**	359.60	**132**	269.60	**82**	179.60	**32**	89.60	**-18**	-0.40
231	447.80	**181**	357.80	**131**	267.80	**81**	177.80	**31**	87.80	**-19**	-2.20
230	446.00	**180**	356.00	**130**	266.00	**80**	176.00	**30**	86.00	**-20**	-4.00
229	444.20	**179**	354.20	**129**	264.20	**79**	174.20	**29**	84.20	**-21**	-5.80
228	442.40	**178**	352.40	**128**	262.40	**78**	172.40	**28**	82.40	**-22**	-7.60
227	440.60	**177**	350.60	**127**	260.60	**77**	170.60	**27**	80.60	**-23**	-9.40
226	438.80	**176**	348.80	**126**	258.80	**76**	168.80	**26**	78.80	**-24**	-11.20
225	437.00	**175**	347.00	**125**	257.00	**75**	167.00	**25**	77.00	**-25**	-13.00
224	435.20	**174**	345.20	**124**	255.20	**74**	165.20	**24**	75.20	**-26**	-14.80
223	433.40	**173**	343.40	**123**	253.40	**73**	163.40	**23**	73.40	**-27**	-16.60
222	431.60	**172**	341.60	**122**	251.60	**72**	161.60	**22**	71.60	**-28**	-18.40
221	429.80	**171**	339.80	**121**	249.80	**71**	159.80	**21**	69.80	**-29**	-20.20
220	428.00	**170**	338.00	**120**	248.00	**70**	158.00	**20**	68.00	**-30**	-22.00
219	426.20	**169**	336.20	**119**	246.20	**69**	156.20	**19**	66.20	**-31**	-23.80
218	424.40	**168**	334.40	**118**	244.40	**68**	154.40	**18**	64.40	**-32**	-25.60
217	422.60	**167**	332.60	**117**	242.60	**67**	152.60	**17**	62.60	**-33**	-27.40
216	420.80	**166**	330.80	**116**	240.80	**66**	150.80	**16**	60.80	**-34**	-29.20
215	419.00	**165**	329.00	**115**	239.00	**65**	149.00	**15**	59.00	**-35**	-31.00
214	417.20	**164**	327.20	**114**	237.20	**64**	147.20	**14**	57.20	**-36**	-32.80
213	415.40	**163**	325.40	**113**	235.40	**63**	145.40	**13**	55.40	**-37**	-34.60
212	413.60	**162**	323.60	**112**	233.60	**62**	143.60	**12**	53.60	**-38**	-36.40
211	411.80	**161**	321.80	**111**	231.80	**61**	141.80	**11**	51.80	**-39**	-38.20
210	410.00	**160**	320.00	**110**	230.00	**60**	140.00	**10**	50.00	**-40**	-40.00
209	408.20	**159**	318.20	**109**	228.20	**59**	138.20	**9**	48.20	**-41**	-41.80
208	406.40	**158**	316.40	**108**	226.40	**58**	136.40	**8**	46.40	**-42**	-43.60
207	404.60	**157**	314.60	**107**	224.60	**57**	134.60	**7**	44.60	**-43**	-45.40
206	402.80	**156**	312.80	**106**	222.80	**56**	132.80	**6**	42.80	**-44**	-47.20
205	401.00	**155**	311.00	**105**	221.00	**55**	131.00	**5**	41.00	**-45**	-49.00
204	399.20	**154**	309.20	**104**	219.20	**54**	129.20	**4**	39.20	**-46**	-50.80
203	397.40	**153**	307.40	**103**	217.40	**53**	127.40	**3**	37.40	**-47**	-52.60
202	395.60	**152**	305.60	**102**	215.60	**52**	125.60	**2**	35.60	**-48**	-54.40
201	393.80	**151**	303.80	**101**	213.80	**51**	123.80	**1**	33.80	**-49**	-56.20

2. Temperature Conversion (Fahrenheit to Celsius)

F^0	C^o	F^0	C^o	F^0	C^o	F^0	C^o	F^0	C^o	F^0	C^o
250	121.11	**200**	93.33	**150**	65.56	**100**	37.78	**50**	10.00	**0**	-17.78
249	120.56	**199**	92.78	**149**	65.00	**99**	37.22	**49**	9.44	**-1**	-18.33
248	120.00	**198**	92.22	**148**	64.44	**98**	36.67	**48**	8.89	**-2**	-18.89
247	119.44	**197**	91.67	**147**	63.89	**97**	36.11	**47**	8.33	**-3**	-19.44
246	118.89	**196**	91.11	**146**	63.33	**96**	35.56	**46**	7.78	**-4**	-20.00
245	118.33	**195**	90.56	**145**	62.78	**95**	35.00	**45**	7.22	**-5**	-20.56
244	117.78	**194**	90.00	**144**	62.22	**94**	34.44	**44**	6.67	**-6**	-21.11
243	117.22	**193**	89.44	**143**	61.67	**93**	33.89	**43**	6.11	**-7**	-21.67
242	116.67	**192**	88.89	**142**	61.11	**92**	33.33	**42**	5.56	**-8**	-22.22
241	116.11	**191**	88.33	**141**	60.56	**91**	32.78	**41**	5.00	**-9**	-22.78
240	115.56	**190**	87.78	**140**	60.00	**90**	32.22	**40**	4.44	**-10**	-23.33
239	115.00	**189**	87.22	**139**	59.44	**89**	31.67	**39**	3.89	**-11**	-23.89
238	114.44	**188**	86.67	**138**	58.89	**88**	31.11	**38**	3.33	**-12**	-24.44
237	113.89	**187**	86.11	**137**	58.33	**87**	30.56	**37**	2.78	**-13**	-25.00
236	113.33	**186**	85.56	**136**	57.78	**86**	30.00	**36**	2.22	**-14**	-25.56
235	112.78	**185**	85.00	**135**	57.22	**85**	29.44	**35**	1.67	**-15**	-26.11
234	112.22	**184**	84.44	**134**	56.67	**84**	28.89	**34**	1.11	**-16**	-26.67
233	111.67	**183**	83.89	**133**	56.11	**83**	28.33	**33**	0.56	**-17**	-27.22
232	111.11	**182**	83.33	**132**	55.56	**82**	27.78	**32**	0.00	**-18**	-27.78
231	110.56	**181**	82.78	**131**	55.00	**81**	27.22	**31**	-0.56	**-19**	-28.33
230	110.00	**180**	82.22	**130**	54.44	**80**	26.67	**30**	-1.11	**-20**	-28.89
229	109.44	**179**	81.67	**129**	53.89	**79**	26.11	**29**	-1.67	**-21**	-29.44
228	108.89	**178**	81.11	**128**	53.33	**78**	25.56	**28**	-2.22	**-22**	-30.00
227	108.33	**177**	80.56	**127**	52.78	**77**	25.00	**27**	-2.78	**-23**	-30.56
226	107.78	**176**	80.00	**126**	52.22	**76**	24.44	**26**	-3.33	**-24**	-31.11
225	107.22	**175**	79.44	**125**	51.67	**75**	23.89	**25**	-3.89	**-25**	-31.67
224	106.67	**174**	78.89	**124**	51.11	**74**	23.33	**24**	-4.44	**-26**	-32.22
223	106.11	**173**	78.33	**123**	50.56	**73**	22.78	**23**	-5.00	**-27**	-32.78
222	105.56	**172**	77.78	**122**	50.00	**72**	22.22	**22**	-5.56	**-28**	-33.33
221	105.00	**171**	77.22	**121**	49.44	**71**	21.67	**21**	-6.11	**-29**	-33.89
220	104.44	**170**	76.67	**120**	48.89	**70**	21.11	**20**	-6.67	**-30**	-34.44
219	103.89	**169**	76.11	**119**	48.33	**69**	20.56	**19**	-7.22	**-31**	-35.00
218	103.33	**168**	75.56	**118**	47.78	**68**	20.00	**18**	-7.78	**-32**	-35.56
217	102.78	**167**	75.00	**117**	47.22	**67**	19.44	**17**	-8.33	**-33**	-36.11
216	102.22	**166**	74.44	**116**	46.67	**66**	18.89	**16**	-8.89	**-34**	-36.67
215	101.67	**165**	73.89	**115**	46.11	**65**	18.33	**15**	-9.44	**-35**	-37.22
214	101.11	**164**	73.33	**114**	45.56	**64**	17.78	**14**	-10.00	**-36**	-37.78
213	100.56	**163**	72.78	**113**	45.00	**63**	17.22	**13**	-10.56	**-37**	-38.33
212	100.00	**162**	72.22	**112**	44.44	**62**	16.67	**12**	-11.11	**-38**	-38.89
211	99.44	**161**	71.67	**111**	43.89	**61**	16.11	**11**	-11.67	**-39**	-39.44
210	98.89	**160**	71.11	**110**	43.33	**60**	15.56	**10**	-12.22	**-40**	-40.00
209	98.33	**159**	70.56	**109**	42.78	**59**	15.00	**9**	-12.78	**-41**	-40.56
208	97.78	**158**	70.00	**108**	42.22	**58**	14.44	**8**	-13.33	**-42**	-41.11
207	97.22	**157**	69.44	**107**	41.67	**57**	13.89	**7**	-13.89	**-43**	-41.67
206	96.67	**156**	68.89	**106**	41.11	**56**	13.33	**6**	-14.44	**-44**	-42.22
205	96.11	**155**	68.33	**105**	40.56	**55**	12.78	**5**	-15.00	**-45**	-42.78
204	95.56	**154**	67.78	**104**	40.00	**54**	12.22	**4**	-15.56	**-46**	-43.33
203	95.00	**153**	67.22	**103**	39.44	**53**	11.67	**3**	-16.11	**-47**	-43.89
202	94.44	**152**	66.67	**102**	38.89	**52**	11.11	**2**	-16.67	**-48**	-44.44
201	93.89	**151**	66.11	**101**	38.33	**51**	10.56	**1**	-17.22	**-49**	-45.00

3. Melting and Boiling Points of Elements

ELEMENT	Sym bol	Melting Point °C.	Boiling point °C.
Actinium	Ac	1050	3200 ± 300
Aluminum	Al	0660.37	2467
Americium	Am	994 ± 4	2607
Antimony	Sb	630.74	1750
Argon	Ar	- 189.2	- 185.7
Arsenic (gray)	As	817 (28 atm)	613 (sub.)
Astatine	At	302	337
Barium	Ba	725	1640
Berkelium	Bk	-	-
Beryllium	Be	1278 ± 5	2970(5 mm)
Bismuth	Bi	271.3	1560 ± 5
Boron	B	2300	2550(sub.)
Bromine	Br	- 7.2	58.78
Cadmium	Cd	320.9	765
Calcium	Ca	839 ± 2	1484
Californium	Cf	-	-
Carbon	C	~ 3550	4827
Cerium	Ce	799 ± 3	3426
Cesium	Cs	28.40 ± 0.01	678.4
Chlorine	Cl	- 100.98	- 34.6
Chromium	Cr	1857 ± 20	2672
Cobalt	Co	1495	2870
Copper	Cu	1083.4 ± 0.2	2567
Curium	Cm	1340 ± 40	-
Dysprosium	Dy	1412	2562
Einsteinium	Es	-	-
Erbium	Er	1529	2863
Europium	Eu	822	1597
Fermium	Fm	-	-
Fluorine	F	- 219.62	- 188.14
Francium	Fr	(27)	(677)
Gadolinium	Gd	1313 ± 1	3266
Gallium	Ga	29.78	2403
Germanium	Ge	937.4	2830
Gold	Au	1064.43	2807
Hafnium	Hf	2227 ± 20	4602
Helium	He	- 272.2 ^{26atm}	- 268.934
Holmium	Ho	1474	2695
Hydrogen	H	- 259.14	- 252.87
Indium	In	156.61	2080
Iodine	I	113.5	184.35
Iridium	Ir	2410	4130
Iron	Fe	1535	2750
Krypton	Kr	- 156.6	- 152.30 ± 0.10
Lanthanum	La	921 ± 5	3457
Lawrencium	Lr	-	-
Lead	Pb	327.502	1740
Lithium	Li	180.54	1347
Lutetium	Lu	1663 ± 5	3395
Magnesium	Mg	648.8 ± 0.5	1090
Manganese	Mn	1244 ± 3	1962
Mendelevium	Md	-	-
Mercury	Hg	- 38.87	356.58
Molybdenum	Mo	2617	4612
Neodymium	Nd	1021	3068
Neon	Ne	- 248.67	- 246.048
Neptunium	Np	640 ± 1	3902
Nickel	Ni	1453	2732
Niobium-Columbium	Nb	2468 ± 10	4742
Nitrogen	N	- 209.86	- 195.8
Nobelium	No		
Osmium	Os	3045 ± 30	5027 ± 100
Oxygen	0	- 218.4	- 182-962
Palladium	Pd	1552	3140
Phosphorus	P	44.1 (white)	280 (white)
Platinum	Pt	1772	3827 ± 100
Plutonium	Pu	641	3232
Polonium	Po	254	962
Potassium	K	63.65	774
Praeseodymium	Pr	931	3512
Promethium	Pm	~1080	2460(?)
Protactinium	Pa	< 1600	-
Radium	Ra	700	1140
Radon	Rn	- 71	- 61.8
Rhenium	Re	3180	5627 (est.)
Rhodium	Rh	1966 ± 3	3727 ± 100
Rubidium	Rb	38.89	688
Ruthenium	Ru	2310	3900
Samarium	Sm	1077	1791
Scandium	Sc	1541	2831
Selenium	Se	217	684.9 ± 1.0
Silicon	Si	1410	2355
Silver	Ag	961.93	2212
Sodium	Na	97.81± 0.03	882.9
Strontium	Sr	769	1384
Sulfur	S	112.8	444.674
Tantalum	Ta	2996	5425 ± 100
Technetium	Tc	2172	4877
Tellurium	Te	449.5 ± 0.3	989.8 ± 3.8
Terbium	Tb	1356	3123
Thallium	Tl	303.5	1457 ± 10
Thorium	Th	1750	~4790
Thulium	Tm	1545 ± 15	1947
Tin	Sn	231.9681	2270
Titanium	Ti	1660 ± 10	3287
Tungsten	W	3410 ± 20	5660
Uranium	U	1132.3 ± 0.8	3818
Vanadium	V	1890 ± 10	3380
Wolfram	(see	Tungsten)	
Xenon	Xe	- 111.9	- 107.1 ± 3
Ytterbium	Yb	819	1194
Yttrium	Y	522	3338
Zinc	Zn	19.58	907
Zirconium	Zr	1852 ± 2	4377

sub. = sublimation

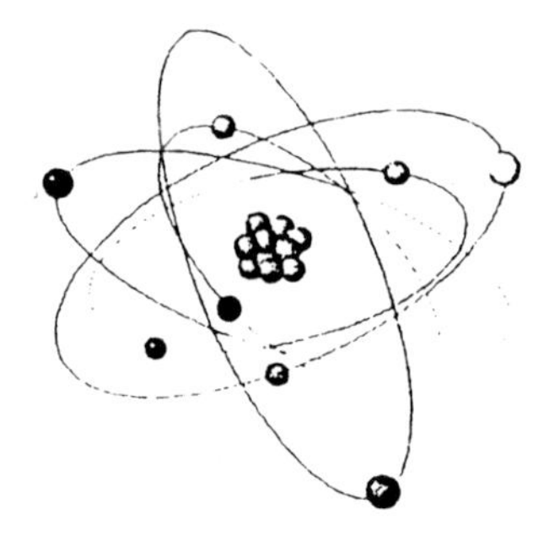

4. Range of Resistances

	ohm-meters at 0°C
Aluminum	2.63×10^{-8}
Amber	5×10^{14}
Boron	1×10^{6}
Constantin	49×10^{-8}
Copper	1.54×10^{-8}
Germanium	2 X 10
Glass	$10^{11} - 10^{13}$
Gold	2.27×10^{-8}
Iron	11.0×10^{-8}
Mercury	94×10^{-8}
Mica (colorless)	2×10^{15}
Nichrome	3.5×10^{-5}
Platinum	11.0×10^{-8}
Quartz (fused)	5×10^{17}
Silicon	3×10^{4}
Silver	1.52×10^{-8}
Sulfur	1×10^{15}
Tungsten	5.0×10^{-8}
Wood (maple)	3×10^{8}

These are only approximations since exact values vary.

5. Coefficients of Volume Expansion

(Increase per unit volume per °C X 10^{-3})

Alcohol (ethyl)	1.1
Benzene	1.2
Carbon tetrachloride	1.2
Ether	1.7
Gases (at 0° C and constant pressure)	3.7
Glass	.025
Glycerin	.5
Mercury	.18
Petroleum	.96
Turpentine	.97
Water (at 20° C)	.21

6. Electrochemical Equivalents

Element and valence	Symbol and charge on ion (valence)	Grams per Coulomb or ampere-second
Aluminum	Al+++	9.32×10^{-5}
Copper (2)	Cu++	32.9×10^{-5}
Copper (1)	Cu+	65.9×10^{-5}
Gold (3)	Au+++	68.1×10^{-5}
Gold (1)	Au+	204.4×10^{-5}
Hydrogen	H+	1.04×10^{-5}
Iron (3)	Fe+++	19.3×10^{-5}
Iron (2)	Fe++	28.9×10^{-5}
Lead (4)	Pb++++	53.7×10^{-5}
Lead (2)	Pb++	107.4×10^{-5}
Nickel	Ni++	30.4×10^{-5}
Oxygen	O^{--}	8.29×10^{-5}
Silver	Ag +	111.8×10^{-5}
Zinc	Zn++	33.9×10^{-5}

7. Wavelengths of Various Radiations

Radiation	Ångström units
Cosmic rays	0.0005
Gamma rays	0.005-1.40
X-rays	0.1-100
Ultraviolet	2920-4000
Visible spectrum	4000-7000
Violet	4000-4240
Blue	4240-4912
Green	4912-5750
Yellow	5750-5850
Orange	5850-6470
Red	6470-7000
Maximum visibility	5560
Infrared	over-7000

Ång. to inch multiply by 3.937×10^{-9}

Ång. to cm.multiply by 1×10^{-8}

8. Specific Heat of Materials

(calories/gram/°C from 20-100°C)

Air (constant pressure)	0.24
Alcohol	0.66
Aluminum	0.22
Cellulose (dry)	0.37
Charcoal (at 10° C)	0.16
Clay (dry)	0.22
Copper	0.09
Glycerin	0.57
Granite	0.19
Gold	0.03
Glass (normal thermometer)	0.20
Ice (at - 10°C)	0.53
I ron (cast)	0.119
Iron (wrought)	0.115
Lead	0.031
Marble	0.21
Mercury	0.033
Paraffin	0.7
Platinum	0.03
Petroleum	0.51
Silver	0.06
Water	1.00
Water vapor (constant pressure)	0.46
Water vapor (constant volume)	0.36

9. Coefficient of Linear Expansion

(approximate expansion per unit length per 0° C x 10^{-6})

Aluminum	24
Brass	19
Bronze	27
Cement	10-14
Copper	1.7
German silver	1.8
Glass (plate)	.9
Gold	13
Granite	8
Ice	51
Iron (wrought)	11
Lead	29
Magnesium	27
Platinum	.9
Pyrex	3.6
Quartz (fused)	.4
Rubber (hard)	80
Silver	1.7
Steel	13
Steel (stainless)	18
Tin	27
Wood (parallel to grain) beech	.2
elm	.6
oak	.5
pine	.5
walnut	.7
Wood (across grain) beech	61
elm	44
oak	
pine	34
walnut	48

10. Density of Liquids

	approx. gm/cm^3 at 20°C
Acetone	0.79
Alcohol (ethyl)	0.79
Alcohol (methyl)	0.81
Benzene	0.90
Carbon disulfide	1.29
Carbon tetrachloride	1.56
Chloroform	1.50
Ether	0.74
Gasoline	0.68
Glycerin	1.26
Kerosene	0.82
Linseed oil (boiled)	0.94
Mercury	13.6
Milk	1.03
Naphtha (petroleum)	0.67
Olive oil	0.92
Sulfuric acid	1.82
Turpentine	0.87
Water0° C	0.99
Water4° C	1.00
Water - sea	1.03

11. Altitude, Barometer, and Boiling Point

altitude (approx. ft)	barometer reading(cm of mercury)	boiling point (° C)
15,430	43.1	84.9
10,320	52.0	89.8
6190	60.5	93.8
5510	62.0	94.4
5060	63.1	94.9
4500	64.4	95.4
3950	65.7	96.0
3500	66.8	96.4
3060	67.9	96.9
2400	69.6	97.6
2060	70.4	97.9
1520	71.8	98.5
970	73.3	99.0
530	74.5	99.5
0	76.0	100.0
- 550	77.5	100.5

12. Specific Gravity

gram /cm^3 at 20° C.

Agate	2.5-2.6	Granite*	2.7	Polystyrene	1.06
Aluminum	2.7	Graphite	2.2	Quartz	2.6
Brass*	8.5	Human body - normal	1.07	Rock salt	2.1-2.2
Butter	0.86	Human body - lungs full	1.00	Rubber (gum)	0.92
Cellural cellulose acetate	0.75	Ice	0.92	Silver	10.5
Celluloid	1.4	Iron (cast)*	7.9	Steel	7.8
Cement*	2.8	Lead	11.3	Sulfur (roll)	2.0
Coal (anthracite)*	1.5	Limestone	2.7	Tin	7.3
Coal (bituminous)*	1.3	Magnesium	1.74	Tungsten	18.8
Copper	8.9	Marble*	2.7	Wood Rock Elm	0.76
Cork	0.22-0.26	Nickel	8.8	Balsa	0.16
Diamond	3.1-3.5	Opal	2.1-2.3	Red Oak	0.67
German Silver	8.4	Osmium	22.5	Southern Pine	0.56
Glass (common)	2.5	Paraffin	0.9	White Pine	0.4
Gold	19.3	Platinum	21.4	Zinc	7.1

*Non homogeneous material. Specific gravity may vary. Table gives average value.

13. Units: Conversions and Constants

FROM	TO	X BY
Acres	Square feet	43560
Acres	Square meters	4046.8564
Acre-feet	Cu-feet	43560
Avogadro's number	6.02252×10^{23}	
Barrel (US dry)	Barrel (US liquid)	0.96969
Barrel (US liq.)	Barrel (US dry)	1.03125
Bars	Atmospheres	0.98692
Bars	Grams/sq.cm.	1019.716
Cubic feet	Acre-feet	2.2956841×10^{-5}
Cubic feet	Cu.centimeters	28316.847
Cubic feet	Cu.meters	0.028316984
Cubic feet	Gallons (US liquid)	7.4805195
Cubic feet	Quarts (US liquid)	29.922078
Cubic inches	Cu. centimeters	16.38706
Cubic inches	Cu. feet	0.0005787037
Cubic inches	Gallons (US liquid)	0.004329004
Cubic inches	Liters	0.016387064
Cubic inches	Ounces (US, fluid)	0.5541125
Cubic inches	Quarts (US. liquid)	0.03463203
Cubic meters	Acre-feet	0.0008107131
Cubic meters	Barrels (US liquid)	8.386414
Cubic meters	Cubic feet	35.314667
Cubic meters	Gallons (US liquid)	264.17205
Cubic meters	Quarts (US Liquid)	1056.6882
Cu. yards	Cu. Cm.	764554.86
Cu. yards	Cu. feet	27
Cu. yards	Cu. inches	46,656
Cu. yards	Liters	764.55486
Cu. yards	Quarts (US Liquid)	807.89610
Days (mean solar)	Days (Sidereal)	1.0027379
Days (mean solar)	Hours (mean solar)	24
Days (mean solar)	Hours (sidereal)	24.065710
Days (mean solar)	Years (Calendar)	0.002739726
Days (mean solar)	Years (sidereal)	0.0027378031
Days (mean solar)	Years (tropical)	0.0027379093
Days (sidereal)	Days (mean solar)	0.99726957
Days (sidereal)	Hours (mean solar)	23.93447
Days (sidereal)	Hours (sidereal)	24
Days (sidereal)	Minutes (mean solar)	1436.0682
Days (sidereal)	Minutes (sidereal)	1440
Days (sidereal)	Second (sidereal)	86400
Days (sidereal)	Years (calendar)	0.0027322454
Days (sidereal)	Years (sidereal)	0.0027303277

13. Units: Conversions and Constants *(continued)*

FROM	TO	X BY
Days (sidereal)	Years (tropical)	0.0027304336
Decibels	Bels	0.1
Decimeters	Feet	0.32808399
Decimeters	Inches	3.9370079
Degrees	Minutes	60
Degrees	Radians	0.017453293
Degrees	Seconds	3600
Dekameters	Feet	32.808399
Dekameters	Inches	393.70079
Dekameters	Yards	10.93613
Decimeters	Feet	0.32808399
Decimeters	Inches	3.9370079
Decimeters	Meters	0.1
Degrees	Circles	0.0027777
Degrees	Minutes	60
Degrees	Quadrants	0.0111111
Degrees	Radians	0.017453293
Degrees	Seconds	600
Dekaliters	Pecks (U.S.)	1.135136
Dekameters	Pints (U.S. dry)	19.16217
Dekameters	Centimeters	1000
Dekameters	Feet	32.808399
Dekameters	Inches	393.70079
Dekameters	Yards	10.93613
Fathoms	Centimeters	182.88
Fathoms	Feet	6
Fathoms	Inches	72
Fathoms	Meters	1.8288
Fathoms	Miles (naut. Int.)	0.00098747300
Fathoms	Miles (statute)	0.001136363
Fathoms	Yards	2
Feet	Centimeters	30.48
Feet	Fathoms	0.166666
Feet	Furlongs	0.00151515
Feet	Inches	12
Feet	Meters	0.3048
Feet	Microns	304800
Feet	Miles (naut. Int.)	0.00016457883
Feet	Miles (statute)	0.000189393
Feet	Rods	0.060606
Feet	Yards	0.333333
Gallons (U.S. liq.).	Acre-feet	3.0688833×10^{-6}

13. Units: Conversions and Constants *(continued)*

FROM	TO	X BY
Gallons (U.S. liq.)	Barrels (U.S. liq.)	0.031746032
Gallons (U.S. liq.)	Bushels (U.S.)	0.10742088
Gallons (U.S. liq.)	Cu. centimeters	3785.4118
Gallons (U.S. liq.)	Cu. feet	0.133680555
Gallons (U.S. liq.)	Cu.inches	231
Gallons (U.S. liq.)	Cu. meters	0.0037854118
Gallons (U.S. liq.)	Cu. yards	0.0049511317
Gallons (U.S. liq.)	Gallons (U.S. dry)	0.85936701
Gallons (U.S. liq.)	Gallons (wine)	1
Gallons (U.S. liq.)	Gills (U.S.)	32
Gallons (U.S. liq.)	Liters	3.7854118
Gallons (U.S. liq.)	Ounces (U.S. fluid)	128
Gallons (U.S. liq.)	Pints (U.S. liq.)	8
Gallons (U.S. liq.)	Quarts (U.S. liq.)	4
Grains	Carats (metric)	0.32399455
Grains	Drams (apoth. or troy)	0.016666
Grains	Drams (avdp.)	0.036671429
Grains	Grams	0.06479891
Grains	Milligrams	64.79891
Grains	Ounces (apoth. or troy)	0.0020833
Grains	Ounces (avdp.)	0.0022857143
Grams	Carats (metric)	5
Grams	Drams (apoth. or troy)	0.25720597
Grams	Drams (avdp.)	0.56438339
Grams	Dynes	980.665
Grams	Grains	15.432358
Grams	Ounces (apoth. or troy)	0.032150737
Grams	Ounces (avdp.)	0.035273962
Gravitational constant	Cm./(sec. X sec.)	980.621
Gravitational constant = G	dyne cm^2 g^{-2}	6.6732(31) X 10^{-8}
Gravitational constant	Ft./(sec. X sec.)	32.1725
Gravitational constant = G	N m^2 kg^{-2}	6.6732(31) X 10^{-11}
Gravity on Earth =1	Gravity on Jupiter	2.305
Gravity on Earth =1	Gravity on Mars	0.3627 Equatorial
Gravity on Earth =1	Gravity on Mercury	0.3648 Equatorial
Gravity on Earth =1	Gravity on Moon	0.1652 Equatorial
Gravity on Earth =1	Gravity on Neptune	1.323 ± 0.210 Equatorial
Gravity on Earth =1	Gravity on Pluto	0.0225 ± 0.217 Equatorial
Gravity on Earth =1	Gravity on Saturn	0.8800 Equatorial
Gravity on Earth =1	Gravity on Sun	27.905 Equatorial
Gravity on Earth =1	Gravity on Uranus	0.9554 ± 0.168

13. Units: Conversions and Constants *(continued)*

FROM	TO	X BY
		Equatorial
Gravity on Earth =1	Gravity on Venus	0.9049 Equatorial
Hectares	Acres	2.4710538
Hectares	Sq. feet	107639.10
Hectares	Sq. meters	10000
Hectares	Sq. miles	0.0038610216
Hectares	Sq. rods	395.36861
Hectograms	Pounds (apoth or troy)	0.26792289
Hectograms	Pounds (avdp.)	0.22046226
Hectoliters	Cu. cm.	1.00028×10^{5}
Hectoliters	Cu. feet	3.531566
Hectoliters	Gallons (U.S. liq.)	26.41794
Hectoliters	Ounces (U.S.) fluid	3381.497
Hectoliters	Pecks (U.S.)	11.35136
Hectometers	Feet	328.08399
Hectometers	Rods	19.883878
Hectometers	Yards	109.3613
Horsepower	Horsepower (electric)	0.999598
Horsepower	Horsepower (metric)	1.01387
Horsepower	Kilowatts	0.745700
Horsepower	Kilowatts (Int.)	0.745577
Horsepower-hours	Kw.-hours	0.745700
Horsepower-hours	Watt-hours	745.700
Hours (mean solar)	Days (mean solar)	0.0416666
Hours (mean solar)	Days (sidereal)	0.041780746
Hours (mean solar)	Hours (sidereal)	1.00273791
Hours (mean solar)	Minutes (mean solar)	60
Hours (mean solar)	Minutes (sidereal)	60.164275
Hours (mean solar)	Seconds (mean solar)	3600
Hours (mean solar)	Seconds (sidereal)	3609.8565
Hours (mean solar)	Weeks (mean calendar)	0.0059523809
Hours (sidereal)	Days (mean solar)	0.41552899
Hours (sidereal)	Days (sidereal)	0.0416666
Hours (sidereal)	Hours (mean solar)	0.99726957
Hours (sidereal)	Minutes (mean solar)	59.836174
Hours (sidereal)	Minutes (sidereal)	60
Inches	Ångström units	2.54×10^{8}
Inches	Centimeters	2.54
Inches	Cubits	0.055555
Inches	Fathoms	0.013888
Inches	Feet	0.083333
Inches	Meters	0.0254

13. Units: Conversions and Constants *(continued)*

FROM	TO	X BY
Inches	Mils	1000
Inches	Yards	0.027777
Kilograms	Drams(apoth. or troy)	257.20597
Kilograms	Drams (avdp.)	564.38339
Kilograms	Dynes	980665
Kilograms	Grains	15432.358
Kilograms	Hundredweights (long)	0.019684131
Kilograms	Hundredweights (short)	0.022046226
Kilograms	Ounces (apoth. or troy)	32.150737
Kilograms	Ounces (avdp.)	35.273962
Kilograms	Pennyweights	643.01493
Kilograms	Pounds (apoth. or troy)	2.6792289
Kilograms	Pounds (avdp.)	2.2046226
Kilograms	Quarters (U.S. long)	0.0039368261
Kilograms	Scruples (apoth.)	771.61792
Kilograms	Tons (long)	0.00098420653
Kilograms	Tons (metric)	0.001
Kilograms	Tons (short)	0.0011023113
Kilograms/cu. meter	Grams/cu. cm.	0.001
Kilograms/cu. meter	Lb. /cu. ft.	0.062427961
Kilograms/cu. meter	Lb./cu. inch	3.6127292×10^{-5}
Kiloliters	Cu. centimeters	1×10^{6}
Kiloliters	Cu. feet	35.31566
Kiloliters	Cu. inches	61025.45
Kiloliters	Cu. meters	1.000028
Kiloliters	Cu. yards	1.307987
Kiloliters	Gallons (U.S. dry)	27.0271
Kiloliters	Gallons (U.S. liq.)	264.1794
Kilometers	Astronomical units	6.68878×10^{-9}
Kilometers	Feet	3280.8399
Kilometers	Light years	1.05702×10^{-13}
Kilometers	Miles (naut. Int.)	0.53995680
Kilometers	Miles (statute)	0.62137119
Kilometers	Rods	198.83878
Kilometers	Yards	1093.6133
Kilometers/hr.	Cm./sec.	27.7777
Kilometers/hr.	Feet/hr.	3280.8399
Kilometers/hr.	Feet/min.	54.680665
Kilometers/hr.	Knots (Int.)	0.53995680
Kilometers/hr.	Meters/sec.	0.277777
Kilometers/hr.	Miles (statute)/hr.	0.62137119
Kilometers/min.	Cm./sec.	1666.666

13. Units: Conversions and Constants *(continued)*

FROM	TO	X BY
Kilometers/min.	Feet/min.	3280.8399
Kilometers/min.	Kilometers/hr.	60
Kilometers/min.	Knots (Int.)	32.397408
Kilometers/min.	Miles/hr.	37.282272
Kilometers/min.	Miles/min.	0.62137119
Kilowatt-hours	Joules	3.6×10^6
Light, velocity of	299,792.4562 Km/sec. ± 1.1	meter/sec. (100x more accurate)
Light, velocity of	m/sec. ± 0.33ppm	$2.9979250(10) \times 10^8$
Light, velocity of	cm/sec. ± 0.33ppm	$2.9979250(10) \times 10^{10}$
Light years	Astronomical units	63279.5
Light years	Kilometers	9.46055×10^{12}
Light years	Miles (statute)	5.87851×10^{12}
Liters	Bushels (U.S.)	0.02837839
Liters	Cu. centimeters	1000
Liters	Cu. feet	0.03531566
Liters	Cu. inches	61.02545
Liters	Cu. meters	0.001
Liters	Cu. yards	0.001307987
Liters	Drams (U.S. fluid)	270.5198
Liters	Gallons (U.S. dry)	0.2270271
Liters	Gallons (U.S. liq.)	0.2641794
Liters	Gills (U.S.)	8.453742
Liters	Hogsheads	0.004193325
Liters	Minims (U.S.)	16231.19
Liters	Ounces (U.S. fluid)	33.81497
Liters	Pecks (U.S.)	0.1135136
Liters	Pints (U.S. dry)	1.816217
Liters	Pints (U.S. liq.)	2.113436
Liters	Quarts (U.S. dry)	0.9081084
Liters	Quarts (U.S. liq.)	1.056718
Liters/min	Cu. ft./min.	0.03531566
Liters/min	Cu. ft./sec.	0.0005885943
Liters/min	Gal. (U.S. liq.)/min.	0.2641794
Liters/sec.	Cu. ft./min.	2.118939
Liters/sec.	Cu. ft./sec.	0.03531566
Liters/sec.	Cu. yards/min.	0.07847923
Liters/sec.	Gal. (U.S. liq.)/min.	15.85077
Liters/sec.	Gal. (U.S. liq.)/sec.	0.2641794
Lumens	Candle power	0.079577472
Meters	Ångström units	1×10^{10}
Meters	Fathoms	0.54680665
Meters	Feet	3.2808399

13. Units: Conversions and Constants *(continued)*

FROM	TO	X BY
Meters	Furlongs	0.0049709695
Meters	Inches	39.370079
Meters	Megameters	1×10^{-6}
Meters	Miles (naut. Int.)	0.00053995680
Meters	Miles (statute)	0.00062137119
Meters	Millimicrons	1×10^{9}
Meters	Mils	39370.079
Meters	Rods	0.19883878
Meters	Yards	1.0936133
Meters/hr.	Feet/hr.	3.2808399
Meters/hr.	Feet/min.	0.054680665
Meters/hr.	Knots (Int.)	0.00053995680
Meters/hr.	Miles (statute)/hr.	0.00062137119
Meters/min.	Cm./sec.	1.666666
Meters/min.	Feet/min.	3.2808399
Meters/min.	Feet/sec.	0.054680665
Meters/min.	Kilometers/hr.	0.06
Meters/min.	Knots (Int.)	0.032397408
Meters/min.	Miles (statute)/hr.	0.037282272
Meters/sec.	Feet/min.	196.85039
Meters/sec.	Feet/sec.	3.2808399
Meters/sec.	Kilometers/hr.	3.6
Meters/sec.	Kilometers/min.	0.06
Meters/sec.	Miles (statute)/hr.	2.2369363
Meter-candles	Lumens/sq. meter	1
Micrograms	Grams	1×10^{-6}
Micrograms	Milligrams	0.001
Micromicrons	Ångström units	0.01
Micromicrons	Centimeters	1×10^{-10}
Micromicrons	Inches	$3.9370079 \times 10^{-11}$
Micromicrons	Meters	1×10^{-12}
Micromicrons	Microns	1×10^{-6}
Microns	Ångström units	10000
Microns	Centimeters	0.0001
Microns	Feet	3.2808399×10^{-6}
Microns	Inches	3.9370070×10^{-5}
Microns	Meters	1×10^{-6}
Microns	Millimeters	0.001
Microns	Millimicrons	1000
Miles (statute)	Centimeters	160934.4
Miles (statute)	Feet	5280
Miles (statute)	Furlongs	8

13. Units: Conversions and Constants *(continued)*

FROM	TO	X BY
Miles (statute)	Inches	63360
Miles (statute)	Kilometers	1.609344
Miles (statute)	Light years	1.70111×10^{-13}
Miles (statute)	Meters	1600.344
Miles (statute)	Miles (naut. Int.)	0.86897624
Miles (statute)	Myriameters	0.1609344
Miles (statute)	Rods	320
Miles (statute)	Yards	1760
Miles/hr.	Cm./sec.	44.704
Miles/hr.	Feet/hr.	5280
Miles/hr.	Feet/min.	88
Miles/hr.	Feet/sec	1.466666
Miles/hr.	Kilometers/hr.	1.609344
Miles/hr.	Knots (Int.)	0.86897624
Miles/hr.	Meters/min.	26.8224
Miles/hr.	Miles/min.	0.0166666
Miles/min.	Cm./sec.	2682.24
Miles/min.	Feet/hr.	316800
Miles/min.	Feet/sec.	88
Miles/min.	Kilometers/min.	1.609344
Miles/min.	Knots (Int.)	52.138574
Miles/min.	Meters/min.	1609.344
Miles/min.	Miles/hr.	60
Mlilligrams	Carats (1877)	0.004871
Mlilligrams	Carats (metric)	0.005
Mlilligrams	Drams (apoth. or troy)	0.00025720597
Mlilligrams	Drams (advp.)	0.00056438339
Milligrams	Grains	0.015432358
Milligrams	Grams	0.001
Milligrams	Ounces (apoth. or troy)	3.2150737×10^{-5}
Milligrams	Ounces (avdp.)	3.5273962×10^{-5}
Milligrams	Pounds (apoth. or troy)	2.6792289×10^{-5}
Milligrams	Pounds(avdp.)	2.2046226×10^{-6}
Milligrams/liter	Grains/gal. (U.S.)	0.05841620
Milligrams/liter	Grams/liter	0.001
Milligrams/liter	Parts/million	1; solvent density = 1
Milligrams/liter	Lb./cu. ft.	6.242621×10^{-5}
Milligrams/mm.	Dynes/cm.	9.80665
Milliliters	Cu. cm.	1
Milliliters	Cu. inches	0.06102545
Milliliters	Drams (U.S. fluid)	0.2705198
Milliliters	Gills (U.S.)	0.008453742

13. Units: Conversions and Constants *(continued)*

FROM	TO	X BY
Milliliters	Minims (U.S.)	16.23119
Milliliters	Ounces (U.S. fluid)	0.03381497
Milliliters	Pints (U.S. liq.)	0.002113436
Millimeters	Ångatröm units	1×10^{7}
Millimeters	Centimeters	0.1
Millimeters	Decimeters	0.01
Millimeters	Dekameters	0.0001
Millimeters	Feet	0.0032808399
Millimeters	Inches	0.039370079
Millimeters	Meters	0.001
Millimeters	Microns	1000
Millimeters	Mils	39.370079
Millimicrons	Ängström units	10
Millimicrons	Centimeters	1×10^{-7}
Millimicrons	Inches	3.9370079×10^{-8}
Millimicrons	Microns	0.001
Millimicrons	Millimeters	1×10^{-6}
Minutes (angular)	Degrees	0.0166666
Minutes (angular)	Quadrants	0.000185185
Minutes (angular)	Radians	0.00029088821
Minutes (angular)	Seconds (angular)	60
Minutes (mean solar)	Days (mean solar)	0.0006944444
Minutes (mean solar)	Days (sidereal)	0.00069634577
Minutes (mean solar)	Hours (mean solar)	0.0166666
Minutes (mean solar)	Hours (sidereal)	0.016732298
Minutes (mean solar)	Minutes (sidereal)	1.00273791
Minutes (sidereal)	Days (mean solar)	0.00069254831
Minutes (sidereal)	Minutes (mean solar)	0.99726957
Minutes (sidereal)	Months (mean calendar)	2.2768712×10^{-5}
Minutes (sidereal)	Seconds (sidereal)	60
Minutes/cm.	Radians/cm.	0.00029088821
Months (lunar)	Days (mean solar)	29.530588
Months (lunar)	Hours (mean solar)	708.73411
Months (lunar)	Minutes (mean solar)	42524.047
Months (lunar)	Second (mean solar)	2.5514428×10^{-5}
Months (lunar)	Weeks (mean calendar)	4.2186554
Months (mean calendar)	Days (mean solar)	30.416666
Months (mean calendar)	Hours (mean solar)	730
Months (mean calendar)	Months (lunar)	1.0300055
Months (mean calendar)	Weeks (mean calendar)	4.3452381
Months (mean calendar)	Years (calendar)	0.08333333
Months (mean calendar)	Years (sidereal)	0.083274845

13. Units: Conversions and Constants *(continued)*

FROM	TO	X BY
Months (mean calendar)	Years (tropical)	0.083278075
Myriagrams	Pounds (avdp.)	22.046226
Ounces (avdp.)	Drams (apoth. or troy)	7.291666
Ounces (avdp.)	Drams (avdp.)	16
Ounces (avdp.)	Grains	437.5
Ounces (avdp.)	Grams	28.349
Ounces (avdp.)	Ounces (apoth. or troy)	0.9114583
Ounces (avdp.)	Pounds (apoth. or troy)	0.075954861
Ounces (avdp.)	Pounds(avdp.)	0.0625
Ounces (U.S. fluid)	Cu. cm.	29.573730
Ounces (U.S. fluid)	Cu.inches	1.8046875
Ounces (U.S. fluid)	Cu. meters	2.9573730×10^{-5}
Ounces (U.S. fluid)	Drams (U.S. fluid)	8
Ounces (U.S. fluid)	Gallons (U.S. dry)	0.0067138047
Ounces (U.S. fluid)	Gallons (U.S. liq.)	0.0078125
Ounces (U.S. fluid)	Gills (U.S.)	0.25
Ounces (U.S. fluid)	Liters	0.029572702
Ounces (U.S. fluid)	Pints (U.S. liq.)	0.0625
Ounces (U.S. fluid)	Quarts (U.S. liq.)	0.03125
Ounces/sq. inch	Dynes/sq. cm.	4309.22
Ounces/sq. inch	Grams/sq. cm.	4.3941849
Ounces/sq. inch	Pounds/sq. ft.	9
Ounces/sq. inch	Pounds/sq. inch	0.0625
Parts/million	Grains/gal. (U.S.)	0.05841620
Parts/million	Grams/liter	0.001
Parts/million	Milligrams/Liter	1
Pints (U.S. dry)	Bushels (U.S.)	0.015625
Pints (U.S. dry)	Cu. cm.	550.61047
Pints (U.S. dry)	Cu. inches	33.6003125
Pints (U.S. dry)	Gallons (U.S. dry)	0.125
Pints (U.S. dry)	Gallons (U.S. liq.)	0.14545590
Pints (U.S. dry)	Liters	0.5505951
Pints (U.S. dry)	Pecks (U.S.)	0.0625
Pints (U.S. dry)	Quarts (U.S. dry)	0.5
Pints (U.S. liq.)	Cu. cm.	473.17647
Pints (U.S. liq.)	Cu. feet	0.016710069
Pints (U.S. liq.)	Cu. inches	28.875
Pints (U.S. liq.)	Cu. yards	0.00061889146
Pints (U.S. liq.)	Drama (U.S. fluid)	128
Pints (U.S. liq.)	Gallons (U.S. liq.)	0.125
Pints (U.S. liq.)	Gills (U.S.)	4
Pints (U.S. liq.)	Liters	0.4731632

13. Units: Conversions and Constants *(continued)*

FROM	TO	X BY
Pints (U.S. liq.)	Milliliters	473.1632
Pints (U.S. liq.)	Minims (U.S.)	7680
Pints (U.S. liq.)	Ounces (U.S. fluid)	16
Pints (U.S. liq.)	Quarts (U.S. liq.)	0.5
Planck's constant	Erg-seconds	6.6255×10^{-27}
Planck's constant	Joule-seconds	6.6255×10^{-34}
Planck's constant	Joule-sec./Avog. No. (chem.)	3.9905×10^{-10}
Pounds (apoth. or troy)	Drams (apoth. or troy)	96
Pounds (apoth. or troy)	Drams (avdp.)	210.65143
Pounds (apoth. or troy)	Grains	5780
Pounds (apoth. or troy)	Grams	373.24172
Pounds (apoth. or troy)	Kilograms	0.37324172
Pounds (apoth. or troy)	Ounces (apoth. or troy)	12
Pounds (apoth. or troy)	Ounces (avdp.)	13.165714
Pounds (apoth. or troy)	Pounds(avdp.)	0.8228571
Pounds (avdp.)	Drams (apoth. or troy)	116.6686
Pounds (avdp.)	Drams (avdp.)	256
Pounds (avdp.)	Grains	7000
Pounds (avdp.)	Grams	453.59237
Pounds (avdp.)	Kilograms	0.45359237
Pounds (avdp.)	Ounces (apoth. or troy)	14.593333
Pounds (avdp.)	Ounces (avdp.)	16
Pounds (avdp.)	Pounds (apoth. or troy)	1.215277
Pounds (avdp.)	Scruples (apoth.)	350
Pounds (avdp.)	Tons (long)	0.00044642857
Pounds (avdp.)	Tons (metric)	0.00045359237
Pounds (avdp.)	Tons (short)	0.0005
Pounds/cu.ft.	Grams/cu. cm.	0.016018463
Pounds/cu.ft.	Kg./cu. meter	16.018463
Pounds/cu. inch	Grams/cu. cm.	27.679905
Pounds/cu. inch	Grams/liter	27.68068
Pounds/cu. inch	Kg./cu. meter	27679.005
Pounds/gal.(U. S.liq.)	Grams/cu. cm.	0.11982643
Pounds/gal.(U. S.liq.)	Pounds/cu. ft.	7.4805195
Pounds/inch	Grams/cm	178.57967
Pounds/inch	Grams/ft.	5443.1084
Pounds/inch	Grams/inch	453.59237
Pounds/inch	Ounces/cm.	6.2992
Pounds/inch	Ounces/inch	16
Pounds/inch	Pounds/meter	39.370079
Pounds/minute	Kilograms/hr.	27.2155422
Pounds/minute	Kilograms/min.	0.45359237

13. Units: Conversions and Constants *(continued)*

FROM	TO	X BY
Pounds on Earth =1	Pounds on Mars	0.3627 Equatorial
Pounds on Earth =1	Pounds on Mercury	0.3648 Equatorial
Pounds on Earth =1	Pounds on Moon	0.1652 Equatorial
Pounds on Earth =1	Pounds on Neptune	1.323 ± 0.210 Equatorial
Pounds on Earth =1	Pounds on Pluto	0.0225 ± 0.217 Equatorial
Pounds on Earth =1	Pounds on Saturn	0.8800 Equatorial
Pounds on Earth =1	Pounds on Sun	27.905 Equatorial
Pounds on Earth =1	Pounds on Uranus	0.9554 ± 0.168 Equatorial
Pounds on Earth =1	Pounds on Venus	0.9049 Equatorial
Pounds/sq. ft.	Atmospheres	0.000472541
Pounds/sq. ft.	Bars	0.000478803
Pounds/sq. ft.	Cm. of Hg (0°C.)	0.0359131
Pounds/sq. ft.	Dynes/sq. cm.	478.803
Pounds/sq. ft.	Ft. of air (1 atm. 60°F.)	13.096
Pounds/sq. ft.	Grams/sq. cm.	0.48824276
Pounds/sq. ft.	Kg./sq. meter	4.8824276
Pounds/sq. ft.	Mm. of Hg (0°C.)	0.369131
Pounds/sq. inch	Atmospheres	0.0680460
Pounds/sq. inch	Bars	0.0689476
Pounds/sq. inch	Dynes/sq. cm.	68947.6
Pounds/sq. inch	Grams/sq. cm.	70.306958
Pounds/sq. inch	Kg./sq. cm.	0.070306958
Pounds/sq. inch	Mm. of Hg (0°C.)	51.7149
Quarts (U.S. dry)	Bushels (U.S.)	0.03125
Quarts (U.S. dry)	Cu. cm.	1101.2209
Quarts (U.S. dry)	Cu. feet	0.038889251
Quarts (U.S. dry	Cu. inches	67.200625
Quarts (U.S. dry)	Gallons (U.S. dry)	0.25
Quarts (U.S. dry)	Gallons (U.S. liq.)	0.29091180
Quarts (U.S. dry)	Liters	1.1011901
Quarts (U.S. dry)	Pecks (U.S.)	0.125
Quarts (U.S. dry)	Pints (U.S. dry)	2
Quarts (U.S. liq.)	Cu.cm.	946.35295
Quarts (U.S. liq.)	Cu. feet	0.033420136
Quarts (U.S. liq.)	Cu. inches	57.75
Quarts (U.S. liq.)	Drams (U.S. fluid)	256
Quarts (U.S. liq.)	Gallons (U.S. dry)	0.21484175
Quarts (U.S. liq.)	Gallons (U.S. liq.)	0.25
Quarts (U.S. liq.)	Gills (U.S.)	8
Quarts (U.S. liq.)	Liters	0.9463264

13. UNITS: CONVERSIONS AND CONSTANTS (continued)

FROM	TO	X BY
Quarts (U.S. liq.)	Ounces (U.S. fluid)	32
Quarts (U.S. liq.)	Pints (U.S. liq.)	2
Quarts (U.S. liq.)	Quarts (U.S. dry)	0.8593670
Quintals (metric)	Grams	100000
Quintals (metric)	Hundredweights (long)	1.9684131
Quintals (metric)	Kilograms	100
Quintals (metric)	Pounds (avdp.)	220.46226
Radians	Circumferences	0.15915494
Radians	Degrees	57.295779
Radians	Minutes	3437.7468
Radians	Quadrants	0.63661977
Radians	Revolutions	0.15915494
Revolutions	Degrees	360
Revolutions	Grades	400
Revolutions	Quadrants	4
Revolutions	Radians	6.2831853
Seconds (angular)	Degrees	0.000277777
Seconds (angular)	Minutes	0.0166666
Seconds (angular)	Radians	4.8481368×10^{-6}
Seconds (mean solar)	Days (mean solar)	1.1574074×10^{-5}
Seconds (mean solar)	Days (sidereal)	1.1605763×10^{-5}
Seconds (mean solar)	Hours (mean solar)	0.0002777777
Seconds (mean solar)	Hours (sidereal)	0.00027853831
Seconds (mean solar)	Minutes (mean solar)	0.0166666
Seconds (mean solar)	Minutes (sidereal)	0.016712298
Seconds (mean solar)	Seconds (sidereal)	1.00273791
Seconds (sidereal)	Days (mean solar)	1.1542472×10^{-5}
Seconds (sidereal)	Days (sidereal)	1.1574074×10^{-5}
Seconds (sidereal)	Hours (mean solar)	0.00027701932
Seconds (sidereal)	Hours (sidereal)	0.000277777
Seconds (sidereal)	Minutes (mean solar)	0.016621159
Seconds (sidereal)	Minutes (sidereal)	0.0166666
Seconds (sidereal)	Seconds (mean solar)	0.09726957
Sq. Centimeters	Sq. decimeters	0.01
Sq. centimeters	Sq. feet	0.0010763910
Sq. Centimeters	Sq. inches	0.15500031
Sq. Centimeters	Sq. meters	0.0001
Sq. centimeters	Sq. mm.	100
Sq. centimeters	Sq. mile	155000.31
Sq. centimeters	Sq. yards	0.00011959900
Sq. decimeters	Sq. cm.	100
Sq. Decimeters	Sq. inches	15.500031

13. Units: Conversions and Constants *(continued)*

FROM	TO	X BY
Sq. dekameters	Acres	0.024710538
Sq. dekameters	Ares	1
Sq. dekameters	Sq. meters	100
Sq. dekameters	Sq. yards	119.59900
Sq. feet	Acres	2.295684×10^{-5}
Sq. feet	Ares	0.0009290304
Sq. feet	Sq. cm.	929.0304
Sq. feet	Sq.inches	144
Sq. feet	Sq. meters	0.09290304
Sq. feet	Sq. miles	3.5870064×10^{-8}
Sq. Feet	Sq. yards	0.111111
Sq. Hectometers	Sq. meters	10000
Sq. inches	Sq. cm.	6.4516
Sq. inches	Sq. decimeters	0.064516
Sq. inches	Sq. feet	0.0069444
Sq. inches	Sq. meters	0.00064516
Sq. inches	Sq. miles	$2.4909767 \times 10^{-10}$
Sq. inches	Sq. mm	645.16
Sq. inches	Sq. mils	1×10^{-6}
Sq. kilometers	Acres	247.10538
Sq. Kilometers	Sq. feet	1.0763010×10^{7}
Sq. Kilometers	Sq. inches	1.5500031×10^{9}
Sq. Kilometers	Sq. meters	1×10^{6}
Sq. Kilometers	Sq. miles	0.38610216
Sq. Kilometers	Sq. yards	1.1959900×10^{6}
Sq. meters	Acres	0.00024710538
Sq. meters	Ares	0.01
Sq. meters	Hectares	0.0001
Sq. meters	Sq. cm	10000
Sq. meters	Sq. feet	10.763910
Sq. meters	Sq. inches	1550.0031
Sq. meters	Sq. kilometers	1×10^{-6}
Sq. meters	Sq. miles	3.8610218×10^{-7}
Sq. meters	Sq. mm	1×10^{6}
Sq. meters	Sq. yards	1.1959900
Sq. miles	Acres	640
Sq. miles	Hectares	258.99881
Sq. miles	Sq. feet	2.7878288×10^{7}
Sq. miles	Sq. kilometers	2.5899881
Sq. miles	Sq. meters	2589988.1
Sq. miles	Sq. rods	102400
Sq. miles	Sq. yards	3.0976×10^{6}

13. Units: Conversions and Constants *(continued)*

FROM	TO	X BY
Sq. millimeters	Sq. cm.	0.01
Sq. millimeters	Sq.inches	0.0015500031
Sq. millimeters	Sq. meters	1×10^{-6}
Sq. yards	Acres	0.00020661157
Sq. yards	Ares	0.0083612736
Sq. yards	Hectares	8.3612736×10^{-5}
Sq. yards	Sq. cm	8361.2736
Sq. yards	Sq. feet	9
Sq. yards	Sq. inches	1296
Sq. yards	Sq. meters	0.83612736
Sq. yards	Sq. miles	$3.228305785 \times 10^{-7}$
Tons (long)	Kilograms	1016.0469
Tons (long)	Ounces (avdp.)	35840
Tons (long)	Pounds (apoth. or troy)	2722.22
Tons (long)	Pounds(avdp.)	2240
Tons (long)	Tons (metric)	1.0160469
Tons (long)	Tons (short)	1.12
Tons (metric)	Dynes	9.80665×10^{8}
Tons (metric)	Grams	1×10^{6}
Tons (metric)	Kilograms	1000
Tons (metric)	Ounces (avdp.)	35273.962
Tons (metric)	Pounds (apoth. or troy)	2679.2289
Tons (metric)	Pounde(avdp.)	2204.6226
Tons (metric)	Tons (long)	0.98420653
Tons (metric)	Tons (short)	1.1023113
Tons (short)	Kilograms	907.18474
Tons (short)	Ounces (avdp.)	32000
Tons (short)	Pounds (apoth. or troy)	2430.555
Tons (short)	Pounds(avdp.)	2000
Tons (short)	Tons (long)	0.89285714
Tons (short)	Tons (metric)	0.90718474
Velocity of light	cm/sec. ± 0.33ppm	$2.9979250(10) \times 10^{10}$
Velocity of light	m/sec. ± 0.33ppm	$2.9979250(10) \times 10^{8}$
Velocity of light (100xmore accurate)	Km/sec. ± 1.1 meter/sec.	$2.997924562 \times 10^{5}$
Volts	Mks. (r or nr) units	1
Volts (Int.)	Volts	1.000330
Volt-seconds	Maxwells	1×10^{8}
Watts	Kilowatts	0.001
Watts (Int.)	Watts	1.000165
Weeks (mean calendar)	Days (mean solar)	7
Weeks (mean calendar)	Days (sidereal)	7.0191654
Weeks (mean calendar)	Hours (mean solar)	168

13. Units: Conversions and Constants *(continued)*

FROM	TO	X BY
Weeks (mean calendar)	Hours (sidereal)	168.45997
Weeks (mean calendar)	Minutes (mean solar)	10080
Weeks (mean calendar)	Minutes (sidereal)	10107.598
Weeks (mean calendar)	Months (lunar)	0.23704235
Weeks (mean calendar)	Months (mean calendar)	0.23013699
Weeks (mean calendar)	Years (calendar)	0.019178082
Weeks (mean calendar)	Years (sidereal)	0.019164622
Weeks (mean calendar)	Years (tropical)	0.019165365
Yards	Centimeters	91.44
Yards	Cubits	2
Yards	Fathoms	0.5
Yards	Feet.	3
Yards	Furlongs	0.00454545
Yards	Inches	36
Yards	Meters	0.9144
Yards	Rods	0.181818
Yards	Spans	4
Years (calendar)	Days (mean solar)	365
Years (calendar)	Hours (mean solar)	8760
Years (calendar)	Minutes (mean solar)	525600
Years (calendar)	Months (lunar)	12.360065
Years (calendar)	Months (mean calendar)	12
Years (calendar)	Seconds (mean solar)	3.1536×10^7
Years (calendar)	Weeks (mean calendar)	52.142857
Years (calendar)	Years (sidereal)	0.99929814
Years (calendar)	Years (tropical)	0.99933690
Years (leap)	Days (mean solar)	366
Years (sidereal)	Days (mean solar)	365.25636
Years (sidereal)	Days (sidereal)	366.25640
Years (sidereal)	Years (calendar)	1.0007024
Years (sidereal)	Years (tropical)	1.0000388
Years (tropical)	Days (mean solar)	365.24219
Years (tropical)	Days (sidereal)	366.24219
Years (tropical)	Hours (mean solar)	8765.8126
Years (tropical)	Hours (sidereal)	8789.8126
Years (tropical)	Months (mean calendar)	12.007963
Years (tropical)	Seconds (mean solar)	3.1556926×10^7
Years (tropical)	Seconds (sidereal)	3.1643326×10^7
Years (tropical)	Weeks (mean calendar)	52.177456
Years (tropical)	Years (Calendar)	1.0006635
Years (tropical)	Years (sidereal)	0.99996121

Index

A

AC (alternating current), 42, 44
acceleration
    by gravity, 63
    definition of, 28
air pressure, 13
alnico magnet, 33
alternating current, 42
ampere, 47
Ampere, A.M., 47
amplitude
    wave, 72
        definition of, 22
Archimedes' principle, 69
atomic energy, 4
audio spectrum, 73
axle, 57

B

battery testing, 46
bends, diver, 66
Bernoulli's principle, 29
bottle, vacuum, 9

C

C.G. (center of gravity), 55
calibration, balance, 64
calorie
    definition of, 51
    food, 51
Celsius, 50
center of gravity (c.g.)
    definition of, 55
change
    physical, 1
        definition of, 1
chemical energy, 4
circuit, electric
    definition of, 44
color rule, 22
compass, magnetic, 35
compression
    sound wave, 74
condensation, 11
conductor, 48
core, electromagnetic, 43
cosmic rays, 20
current, electric, 47

D

DC (direct current), 44
deafness, 71
decibel meter, 71
decibels, 71
density, 2
    definition of, 67
depth distance, 17
diffraction grating, 25
drag, 63
    definition of, 54
DVM (digital voltmeter), 47

E

efficiency, machine, 61
effort arm, 56
Einstein, Albert, 64
electric
    circuit, 44
    current, 47
    parallel, 45
    resistance to, 48
    series, 44
    short, 46
electrical energy, 4
electricity
    definition of, 37
    static, 37
electromagnet, 40
electromagnetic

force, 52
induction, 43
spectrum, 20
waves, 20
energy
definition of, 53
forms, 4
gravitational, 26
kinetic, 26
potential, 26
waves, 3
equilibrium
forces, 31
pressure, 12
expansion of matter, heat, 6

F

Fahrenheit, 50
field, magnetic, 34
filters, color, 22
food, calorie, 51
force
definition of, 52
frequency, 73
definition of, 21
natural, 75
friction, 28, 61
definition of, 54

G

Galileo, 63
gamma radiation, 20
gravity, 52
center of, 55
formula, 62

H

heat, 20
and expansion of matter, 6
energy, 4, 8
fusion of, 3
radiation, 9
specific, 2
hertz (Hz), 21
Hertz, Heinrich, 21
holograms, 23

I

image
diffused, 16
regular, 16
incidence, angle
light and, 16
inclined plane, 59
incoherent light, 23
induction
electrical, 42
infrared
rays, 8, 20
wavelength, 22
insulator, 48

K

Kelvin, 50
kinetic energy, 26
molecular theory, 10

L

laser, 23
lenses, 24
lever
effort, 56
first-class, 58
fulcrum, 56
resistance, 56
light
behavior of, 14
definition of, 25
energy, 4
incoherent, 23
laser, 23
normal line of, 16
properties of, 23
refraction, 24
visible, 20
linear expansivity, 3
loudness, sound, 71

M

M.A. (mechanical advantage)
actual, 61
ideal, 61
machine
definition of, 56
lever, 56
Magdeburg hemispheres, 12
magnet
alnico, 33
permanent, 33

magnetic
  compass, 35
  field, 34
  force lines, 34
  induction, 42
  of wire, 39
  poles, 32
  strength, increasing, 41
magnetization, of objects, 33
magnets, 32
manometer, 66
mass, 64
matter
  phases of, 10
  physical changes of, 1
  properties of, 2
mechanical advantage (M.A.)
  actual, 61
  definition of, 56
  ideal, 61
mechanical energy, 4
medium, 24
microwaves, 20
molecules, kinetic theory of, 10
moment, 58
motion pictures, 19

N

natural frequency, 75
Newton, 53
Newton, Isaac, 62
Newton's third law, 31
nichrome, 48
nuclear force, 52

O

Oersted, Hans Christian, 39
Ohm, Georg, 48
Ohm's law, 49

P

parallel circuit, 45
persistence, of vision, 19
phases, (states) of matter, 10
photons, 20
pinhole camera, 15
poles
  magnetic, 32
potential energy, 27
potential gravitational energy, 26
pressure, 65
  measuring with manometer, 66
primary coil, 43
prism, 25
pulleys, 60

R

radiation, heat, 9
radio waves, 20
rarefaction, of sound waves, 74
reflection, angle of light, 16
refraction
  laws, 18
  light, 25
resistance
  arm, 56
  electrical, 48
resistivity, 2
resonance, 75

S

schlieren, 2, 5
secondary coil, 43
series circuit, 44
short circuit, electrical, 46
sound, 70–71
  loudness and, 71
  pure, 75
  vibrations and, 4
specific gravity, 2, 67
specific heat, 2
spectrum, electromagnetic, 20
speed, 27
states of matter, 10
static electricity, 37
superconductors, 48

T

temperature, 50
terminal velocity, 63
thermal conductivity, 3
thermometer, 50
torque, 58
transformer, 43
translucency, 22
transparency, 22
transverse waves
  sound of, 74
tungsten, 48

U

ultraviolet light, 22
ultraviolet rays, 20

V

vacuum bottle, 9
Van de Graaff generator, 38
velocity, 27
visible light, 22
vision, persistence of, 19
volt, 47
Volta, Alessandro, 47
voltmeter, 47

W

waves, 21
    amplitude of, 22
    electromagnetic, 20
    energy, 3
weight, 52
    definition of, 64
wheel, 57
work, 53

X

X rays, 20

Y

Young's modulus, 2